HARVESTS AND HARVESTERS

HARVESTS AND HARVESTERS

Fruit and Vegetable Growing in Britain

by

JOHN HARGREAVES

LONDON
VICTOR GOLLANCZ LTD
1987

First published in Great Britain 1987
by Victor Gollancz Ltd,
14 Henrietta Street, London WC2E 8QJ

The quotations from Dr Redcliffe Salaman's
The History and Social Influence of the Potato
are published by kind permission of Cambridge
University Press.

British Library Cataloguing in Publication Data
Hargreaves, John *1950–*
    Harvests and harvesters: fruit and vegetable
growing in Britain.
    1. Fruit-culture—Great Britain—History
    2. Vegetables—Great Britain—History
    I. Title
    634'.0941   SB354.6.G7

ISBN 0-575-03682-6

Typeset at The Spartan Press Ltd,
Lymington, Hants
and printed in Great Britain by St Edmundsbury Press Ltd,
Bury St Edmunds, Suffolk

To
Cyril Hargreaves
and
Audrey Hargreaves

Contents

Acknowledgements

I would like to thank all the growers and their associates who helped me in the preparation of the book, including those who are mentioned by name in the text and the many others who are not.

The short quotation from P. G. Wodehouse (p. 41) on asparagus is from *The Code of the Woosters* by P. G. Wodehouse, Barry and Jenkins (1962), and (p. 137) on glasshouses, from *Wodehouse at Work to the End* by Richard Usborne, Jenkins (1961). The quotation (pp. 15–16) from R. F. Delderfield on Covent Garden appeared in *The Grower* of 19 December 1959. "The Watercress Girl" (p. 27) appears in a collection of stories entitled *Fishmonger's Fiddle* by A. E. Coppard, Cape (1925). I would like to thank Cambridge University Press for permission to quote from *The History and Social Influence of the Potato* (pp. 170–71) by Dr Redcliffe Salaman (1945).

The following books were also consulted, and may be of further interest to the reader:

Prodfact by Daphne MacCarthy, British Farm Produce Council (1986)
Plants in the Service of Man by Edward Hyams, Dent (1971)
Food in England by Dorothy Hartley, MacDonald & Jane (1954)
The Worm Forgives the Plough by John Stewart Collis, C. Knight (1973)
Cultivated Fruits of Britain by F. A. Roach, Blackwell (1985)
English Earth by Marjorie Hessel Tiltman, Harrap (1935)
The Fruit Garden by George Bunyard and Owen Thomas, Country Life (1904)
Grow Your Own by Lawrence Hills, Faber (1971)
The Sandy I Knew by Ken Quince, K. Quince, 166A St Neots Road, Sandy, Beds (1984)
Strawberries in the Wisbech District — The First Seventy Years by Harold C. Selby (copy in the Wisbech District Library)
The Oxford Book of Food Plants by Masefield, Wallis and Harrison, OUP (1969)

Plant and Planet by Anthony Huxley, Allen Lane (1974)
Food by Waverley Root, Simon & Schuster (1980)
Food in History by Reay Tannahill, Eyre Methuen (1973)
The Compleat Strawberry by Stafford Whiteaker, Century (1985)
Garlic by John Blackwood and Stephen Fulder, Javelin (1986)

A note about weights and measures
Most official compilations of horticultural statistics are now expressed in metric units, and a few growers use these exclusively. The majority of growers I visited, however, still use traditional British units, and as much of my material is in their direct speech, this inevitably became my prior mode. For the sake of easy comparison, with a few exceptions, I have therefore converted hectares to acres, kilograms to pounds and tonnes to tons where they appear in my text.

J.H.

HARVESTS AND HARVESTERS

I

Introduction

It WAS NOT a Briton who described patriotism as a love of the good things we ate in our childhood. But a British gardening organization has put together a catalogue of 'Heritage Seeds' which they believe constitute a glorious and significant part of our national inheritance.

Dozens of these outstanding vegetable varieties date from the Victorian and Edwardian eras. During this period the head gardener was a very important and highly skilled member of staff at the country homes of the wealthy. To feed the family and its huge entourage, he organized the year-round provision of vegetables and fruit, often including peaches and nectarines, apricots, grapes, figs, pineapples and melons from hothouses and sheltered spaces within the walled garden. It was also a time of great enthusiasm for amateur gardening among the working class, when leek societies flourished in the West Midlands, for example, and gooseberry clubs in industrial Lancashire. And despite long periods of depression in agriculture, there was a huge expansion of market gardening in many districts to serve the growing cities. It was towards the end of this time, during the First World War, that my grandfather developed nystagmus while digging coal at a Yorkshire mine. Leaving the pit for the last time he spent his entire wages at a nursery, buying five shillings' worth of wallflowers. He sold them in small bunches as he walked, and arrived home with seven shillings and sixpence.

By the time I was born, in 1950, my grandfather had become the head of a large family business of wholesale fruit and potato merchants, with several branches in Yorkshire and Lancashire. Wholesale merchants then dominated the trade in fruit, flowers and vegetables to an extent that they never had before or have since. In 1957, 70 per cent of British horticultural products passed through only 30 primary wholesale markets, and the largest of these, Covent Garden, had a turnover eight times greater than any other. I grew up on these markets. Most of my Saturdays and school holidays were spent at the warehouse, or about its business. I spent hours in the huge empties' yard, helping to sort and stack bushel boxes, half-bushels, flat lettuce boxes, stubby tomato boxes, cabbage crates, strawberry trays, hessian potato sacks . . . each stamped with the grower's name for eventual return. Then I would ride with the lorry drivers out to the farms and was able gradually to put farmers' faces to their names on crates and boxes. I visited glasshouse nurserymen, market gardeners, mushroom growers; large-scale arable farmers, and smallholders with a few plum trees beside the road.

I learned that these growers were as different as their varied skills, and soils, and climates. Some would usher me into a cosy farmhouse, issue a large mug of tea, and make sure that I had a prize specimen of whatever commodity had just come into season, to take home specially for my family. Others would assail me, as the boss's son, with a litany of their problems and expenses, and grumble profoundly about the prices last returned. Both the growers who seemed to me to be in harmony with nature and those who struggled against it had a notable capacity for adopting, like Hoffnung's members of the orchestra, pertinent characteristics of the products of their labour. There were bold, clear-skinned onions; blustering red beets; hearty, shining lettuce heads. And then there were those whose faces, clothes and manner gave nothing away. These were the technocrats who, I supposed, had dispensed with nature altogether. In that case my role was to ferry bits of paper from office to office, collecting the correct data and a number of signatures.

Whoever had grown it, there was the produce itself: sprouts in green nets and carrots in red for a blaze of colour; tomatoes in pink and white tissue paper; lettuces crowning their boxes, and cabbage crates bulging with bellies of greens. I learned all the geometries of making up the load, as the driver would lift the boxes up on to the lorry and I would stack them against the headboard. Bags of potatoes would come up a conveyor from the winter store and we would deck them on straw, ten across and six high, working backwards, with an extra rank on the end to make 10 tons. Sometimes we would be directed out into the fields, if the consignment was not ready, to find a crew cutting lettuce straight into boxes. We would wait, or help with the job, amid the intoxicating smells of ripe fruit, fresh-pulled roots or newly cut greens, and I often felt extraordinarily privileged then to spend so much time, almost all year round, at the point of harvest. It was not always the golden autumnal harvest of the hymn books and farm feast. It was more often wet and cold, and there was invariably a vast amount of mud underfoot. But the memory of the bloom of produce at harvest has stayed with me since childhood.

We went to these farms because they produced what we needed to supply our customers — and they always had produced it, in my brief experience. But as I grew older I was puzzled, and excited, by patterns and logic in this production which seemed as mysterious as they were efficient. Why did we always drive for our forced rhubarb to the heart of industrial Yorkshire? When severe frost struck our cauliflower grower in Lincolnshire, how did my father know to get in touch at once with old associates in the Vale of Evesham? And when I was a step removed from the harvest itself, the geography of it became even more intriguing. In the railway goods' yard, for instance, we unloaded raspberries which came every summer from Blairgowrie, north of Edinburgh; broccoli and daffodils every spring from Mounts Bay in the southern part of Cornwall. We unloaded Manx swedes from a little tramp of a boat at Glasson Dock. And when potatoes were very short we located a pilot who still knew the sands of the Lune estuary, to guide four boats full of Ulster 'praties' — including blue-skinned ones — up the river to the old port of Lancaster, which was my father's base at the time. Here was the entrepreneurial side of things. For as well as the horticultural factors, and the traditions and expertise of the growers, there was always the market, with its unique imperatives. No business could move so fast, I thought, or be so fickle. And however great the errors of judgement, in this trade there could never be any remaindered stock. The great 'bloom' of the harvest had a very short life, and the best must be made of it.

This was illustrated nowhere better than in the speed and bustle of Covent Garden, where according to *Punch* as long ago as 1882, "the purchase and sale of vegetables involves almost as much noise as a French Revolution". Until I became an independent traveller in my late teens, the most I saw of London was the 30 acres between St Martin's Lane, Long Acre, Drury Lane and the Strand. My awareness of our capital city's architectural glory was limited to the Floral Hall, the Charter, Tin and Jubilee Markets, and the portico of St Paul's Church, as my father steered me, wide-eyed at four o'clock in the morning, through the press of buyers, sellers, porters, vehicles of every kind, and ever-shifting piles of produce from every growing district in Britain. As R. F. Delderfield wrote in an article in *The Grower*:

This is the Clapham Junction and the Crewe of half the tomatoes, cabbage, carrots and cauliflowers grown in the British Isles. They [the people of Covent Garden] are members of a tribe that has somehow wandered away from the rest of us and built a strictly private fortress in the streets between the Strand and Holborn. Here they have always lived, speaking their own language and practising their own customs. One can be *with* them but never *of* them. One is either born into their midst or remains outside it, stupefied and amazed at their application to a calling that is almost a cult.

Covent Garden began as a Convent garden belonging to the Abbey of Westminster, and it was the monasteries — which usually excelled at horticulture and often sold surpluses at town markets — which first recorded the development of horticulture in Britain in any detail. Our Iron-Age, Bronze-Age and even Neolithic ancestors cultivated cereals and perhaps a few radish-type roots, although the impression of a Greek geographer at the time of the birth of Christ was that the British "have no experience in gardening or other agricultural pursuits". There may have been some attempts by the Celts to grow tree fruits near their homesteads, but it was the Romans who first laid out large systematic orchards in Britain, introducing many varieties of fruit which had been improved by selecting superior types over generations. They also planted herb gardens, vineyards, and several new vegetables such as turnips. Anglo-Saxon villagers grew turnips, radishes, onions and leeks, undeveloped forms of carrot and parsnip, cabbage, spinach, peas and field beans for their own consumption, but although some fruit trees and their seedlings no doubt survived through the Dark Ages, organized horticulture on the Roman scale ceased to exist. It was revived by the monasteries from AD 600 onwards until their dissolution almost a thousand years later. The monks were progressive farmers, and through their connections with continental Europe were able to introduce many new varieties of fruit and vegetables. As well as the familiar roots and field greens they grew many "herbs for potage and herbs for salad", and cultivated, where practical, apples and pears, quinces, nuts, cherries, grapes and peaches.

During the twelfth and thirteenth centuries farming in general prospered, and every manor and most farms possessed considerable fruit and vegetable gardens. But there followed a long period of severe depression. There was a series of droughts and pestilences, including the Black Death which killed a third of the population, and a gradual deterioration of climate, so that the very cold winters and late springs devastated many vineyards and orchards. Vegetable production also suffered. William Harrison, Dean of Windsor, wrote in 1577: "In the time of Henry IV the decline had so far advanced, that in the reigns of

Henry VII and VIII there was little or no use of them [vegetables] in England, but they remained either unknown or supposed as food more meet for hogs and savage beasts to feed upon, than mankind."

Catherine of Aragon imported a Flemish gardener to raise her salads, and Catherine Parr imported the vegetables themselves. But during Henry VIII's reign large new orchards were planted in Kent, and towards the end of the Tudor period a kitchen garden again became a feature of wealthy houses, often sporting novelty vegetables from the New World. Agriculture was now widely practised as a trade, rather than being simply a subsistence culture. Increasing amounts of cultivated land were let for cash rents and their tenants grew cash crops for sale in the trading towns. A district depended primarily on locally grown supplies, but certain crops began to be associated with certain areas where conditions were favourable and expertise developed. Around 1570, for example, a group of Walloon refugees settled near Sandwich in Kent and began a market garden business which flourished. This was followed shortly by Dutch-sponsored market gardens to the northeast of London. These immigrants from the Low Countries were skilled at draining land, and brought with them new gardening techniques and new varieties of vegetables. The produce was sold at many street and courtyard locations in the capital, and as the city expanded westwards, at Covent Garden.

During the following two centuries an almost constant stream of herbals and gardening books was published, and horticulture became more widely popular. A profession of gardeners with great skills developed in Britain. In 1782, three years into the siege of Gibraltar, the commander of the Spanish troops sent General Eliot a basket of fruit and vegetables for his own refreshment, and he answered: "I entreat your Excellency not to heap any more favours on me of this kind. . . . The English are naturally fond of gardening and cultivation; and here we find our amusement in it, during the intervals of rest from public duty."

The agricultural revolution of the eighteenth century involved more sophisticated crop rotations, selective livestock breeding and the introduction of various new farming implements, but these were concerned only peripherally with fruit and vegetable production. In fact, these improvements necessitated a hugely increased rate of land enclosure which deprived many rural people of the small patches of ground on which they had been able to grow a few root crops and other vegetables. During the Napoleonic War the founder of the first semi-official Board of Agriculture, Sir John Sinclair, urged the nation to "not be satisfied with the liberation of Egypt, or the subjugation of Malta, but let us subdue Finchley Common; let us conquer Hounslow Heath; let us compel Epping Forest to submit to the yoke of improvement". But there was no move towards intensive vegetable production until the practice arose of allotting industrial workers a small potato patch in lieu of part of their wages.

With fruit and vegetables transport was always critical, for as long as this depended on the horse and cart taking produce to market in the morning and bringing horse manure back in the evening, growing was limited to the vicinity of the towns. Less perishable goods could travel by coastal, river and, eventually, canal barge. Following the building of the railways, small market garden areas such as those in Bedfordshire and Worcestershire became important nationally, and expanded rapidly to supply the growing populations of distant cities. Even during prolonged periods of agricultural depression, horticulture gradually expanded. Then, after the First World War, the motor lorry and the tractor led to further change in the structure and location of production. Bulky vegetables such as potatoes, carrots, cabbages and cauliflowers could be grown on land which suited them well, in Lincolnshire and East Anglia, even though far from market or railway station. The growth of the canning and, subsequently, the freezing industry, supported this change. After the Second World War, with far-reaching technical improvements which included the mechanization of many farming processes and the chemical control of weeds, pests and diseases, this move was greatly accelerated. Specialist growers concentrated on one or two vegetables which they grew on a large scale on arable farms. The market gardeners, where they have survived at all, have done so by concentrating on the more valuable or perishable vegetables, especially salad crops brought on early by the use of glass, or by developing a Pick-Your-Own business.

Originally market gardeners carted their own produce to town and sold from their own stands to retailers or consumers. After the arrival of the railways, those located far from the markets entrusted their produce to an agent who sold it for them and took a commission. These agents became the wholesalers, such as my grandfather and father, who also then handled imports. When arable farmers began marketing large quantities of fruit and vegetables they used the same system, but increasingly began asking their own price for their produce, rather than selling on commission. With the changing trend at retail level towards large supermarket chains or 'multiple' stores, some of these large-scale growers were able to bypass the wholesale markets altogether and sell direct to the retailer.

This much, or little, I had learned during my childhood on the markets and in school. Now, as a keen gardener and cook I wanted to flesh-out this skeleton. I wanted to see the fruits and vegetables again in the field and orchard, to enlarge my appreciation of where they come from and why, and to talk to the men and women who grow and harvest them. For there would be many new areas of change: the choice of variety dictated by the latest harvesting and storage techniques; the effects of EEC policies decided by representatives of ten countries, on the other side of the Channel; the demands of a travelled and affluent consumer for wider choice; medical recommend-ations for a national diet higher in fibre, with more fresh produce; the discovery of potentially harmful residues in many fruits and vegetables from chemicals applied during their cultivation, and the growth of an influential 'organic' movement. These were only a few of the issues I expected to arise. Many of the changes would be inevitable, while others would require decisions — from growers, from bureaucrats and politicians, and from increasingly knowledgeable consumers.

2

The Big Stick — Wakefield Rhubarb

THE TINY TRIANGLE of land between Wakefield, Leeds and Dews-bury, in the heart of industrial Yorkshire, is dotted with long, windowless brick sheds. Solitary and weather-beaten at the edge of the fields, these sheds could easily be mistaken by the casual visitor for the soulless outbuildings of old woollen mills, or gloomy remnants of colliery pitheads. Lying empty through the summer and autumn, the doors are sealed to the world during the winter months, and the sheds are a mystery even to many of the local people. "I used to skip past them every day on my way to school," a journalist has written. "And I hadn't a clue what went on inside until I was assigned the story."

What goes on inside has not changed much since the early years of Queen Victoria's reign, when the young monarch was "amused" by the new fashion for forced rhubarb. Long red sticks of Victoria and Prince Albert still emerge from these darkened sheds in the depths of winter, although the number of sheds in use has dropped dramatically over the last twenty years.

When I first peeped, as a young boy, through the small door of one of these singular buildings, I was impressed by the huge scale of the operation. Rhubarb roots weighing up to 60 lb. each were brought in from the field and packed by hand, 5,000 to the shed. The stems were forced to a height of some three feet, at a rate of up to two inches a day. As the crowns of new growth swelled to the size of goose eggs, you could actually hear them popping open and stretching upwards.

Now, as a six-foot tall adult, I am struck by the compactness and constrictions of procedures which conjure up a world before the Factory Acts. The low roofs, barely eight feet high at the centre, slope rapidly away to less than four feet at the edge, to avoid heating unnecessary space, and the roots are packed close together for the same reason. The central path is a foot wide, those leading from it only six inches, and all these paths are ankle-deep in water. The pickers must bend to pull each stick at the root, working in the poorest visibility, for even a little light soon affects the colour of the sticks and their tiny yellow leaves. Many pickers still use candles. "They say old 'annah wore one leg shorter than t'other, picking rhubarb," a retired worker told me. "They say it's growed i' this district 'cause Yorkshiremen are strong in t'back and weak in th' 'ead."

They also say that rhubarb is grown here because the Yorkshireman is especially enterprising: "Dairymen have cows to look after all year round, but these here market gardeners had nowt to do in winter, so they thought to theirselves, 'We'll put up a little shed'. . . ." Or if you ask another man, he'll tell you how rhubarb is a heavy feeder, and that here they've got all the cheap shoddy from the mills, to dig in to the soil and slowly release its nitrogen. And then, mindful perhaps of the few crowns of rhubarb which always seem to thrive in the muckiest little corners of people's gardens: "It feeds off the grime up here, and the pollution," as though the West Riding were the rubbish heap in the garden of England.

Norman Asquith is Chairman of the Leeds and District Market Gardeners and Rhubarb Growers Association. He is a short balding man who grows 100 tons of rhubarb each year and transports the produce of over 30 other growers in his area. I met him in the office of his large packhouse and distribution centre at Brandy Carr Lane, Wakefield. He warmed quickly and enjoyed talking about his crop's "luvly colour" and its "grand taste" in a generous gush of anecdotes and reminiscences. He gave me the most convincing explanation of why 90 per cent of the early rhubarb grown in Britain comes from this little corner of Yorkshire.

"In the West Riding, funnily enough, it gets cooler quicker than anywhere else in the country. By that I mean that at the back end of the year, the soil itself comes down quicker than anywhere else. We want that cold temperature early, to send the rhubarb dormant. We have a friend in Scotland who grows rhubarb, and he can't put his roots in his shed until a month after us. If he put them in the same time we did, he'd have no crop." But he agreed that Yorkshire character came into it as well. "'Course, it keeps you in gainful employment. There's nowt nicer than to come out on a cold winter's morning and go into the warmth and comfort of a rhubarb shed, where you can work all day without your coat on. It's enjoyable work. Mind, I think rhubarb growers are born, not made. Even when I were young . . . and I've been buggering about with rhubarb since I was fourteen, bar six years in the war when I was at t'pit . . . we'd an old stove in the tying shed, and you used to get a lot of comfort in there on a night, when you'd finished tidying up and thinking about what you were doing. Everyone seemed to be 'appy. It were a job they seemed to like doing. If you'd had a month filling t'sheds in t'cold and wet, once your sheds were full you'd something to look forward to. You could go to work and work all day and not be frozen daft, and to me that's a great thing.

"We have to do a lot of promotion," he said, "because folk get lazy and they stop buying rhubarb. In the old days, some growers would have 50 or 60 sheds. They'd have pickers on day shift and night shift, and they'd send a special train from Wakefield to London every day, loaded with nothing but rhubarb. They'd move thousands of tons. Now we send about 50 tonnes a week from here to London. It's getting to be like an exotic. You've got to keep telling the customer what to do with it." He described his latest gimmick: "We've issued an international challenge . . . to anyone in the world who believes they can make a better tasting rhubarb pie than the Yorkshire pies at the Leeds show. We've challenged the Russians! It's supposed to have come from Russia in t'first place, but I know we can grow it redder than t'Reds!"

Our rhubarb, *Rheum rhaponticum*, is generally believed to be descended from a wild Siberian species. It has long been cultivated in the northern provinces of China, and was mentioned in the herbal *Pen-king*, which dates from 2700 BC. The roots were kiln-dried and powdered to provide a purgative medicine which was known to the Greeks and Romans and was still brought to Europe in the Middle Ages along the old spice route. Live roots for propagation were brought to Britain in the sixteenth century and were long associated with the wide valley of the river Volga. *Rha* was the ancient name for the Volga and *barbarum* means *foreign*, giving us the Latin *Rhabarbarum*, or rhubarb.

Rhubarb remained a horticultural curiosity for many years, although some scullery maids learned how to boil up the leaves to clean dirty old pans. (The oxalic acid content of the leaves can be so high as to render them lethal if eaten.) In the eighteenth century the Bath Society for the Encouragement of Agriculture inspired a more widespread cultivation, and the long leaf stems of improved varieties, which are strictly speaking 'vegetables', began to find their way into pies and puddings as 'fruit'.

In the early nineteenth century the technique of forcing with heat and darkness was developed commercially and indoor rhubarb rapidly became very popular. Here was an especially tender and tasty stewing fruit which brought a blaze of fresh colour to a season where fruit of any kind was scarce, although the British were, and still are, the only people to eat rhubarb in any quantity. Contemporary cookery books listed recipes for Imperial Rhubarb Pie, Albert's Fresh Rhubarb Cake, Highland Rhubarb Crisp, Rhubarb Chutney, Rhubarb and Beer Pie, and many others. And the wagons rolled from Wakefield to Covent Garden.

The daughter of a market gardener in the village of Timperley, Cheshire, wrote to me with an interesting footnote to this story. In November 1938 her father, Frank Marsland, was walking through his acre of dormant rhubarb plants and noticed one odd crown which appeared ready to begin growth. He carefully nurtured and propagated this single mutated crown and was soon sending regular supplies of good, strong and very profitable rhubarb to Manchester market before Christmas. After several years, an entire row of these valuable crowns was lifted and stolen away in the dead of night by a stymied rival. Then towards the end of the war Marsland auctioned a number of crowns for the benefit of the Red Cross, and the genie was truly out of the bottle. The original market garden was subsequently sold for housing development, and a school was built on the rhubarb site. But the variety Timperley is now the main early crop for virtually every grower in the country.

From Norman Asquith I drove ten miles to the seven rhubarb sheds of Ken Oldroyd at Ashfield House in the village of Carlton. Mr Oldroyd, a big Yorkshireman, was at work in the muddy yard in his wellies. He spoke slowly, with consideration, but it was evident at once that here was another rhubarb grower born, not made.

"I've got two of the old–type of sheds, but all the others are multi-span which means you've more room to move about inside. They've been re-roofed in the last ten years, using second-hand timber. It used to be cheap, but now second-hand timber is nearly as dear as new. Then three years ago we insulated all the roofs. Fuel is terribly expensive. We used to have a coal fire at each end of the shed, fed by hand, and there'd be a flue going half-way up the shed from each end, full of hot air. Then when oil was about five pence a gallon, we converted. It's still the same idea, with warm air blown by a fan over a flame, but using oil or propane instead of coal.

"We start off by splitting our own roots. You try to maintain the strain, but if you're not careful you soon get a lot of rogues creeping in, so every now and then we bring in new virus-free stocks. We bought some last summer. We've started this new system of rapid propagation in pots which the Ministry is advocating. We split them all through the summer, probably three times, getting two roots from one, on average, every eight weeks. Then we'll keep back some for splitting next year and plant out the surplus, in September. They'll stay in the soil for two years, so it takes about 70 acres just to keep these seven sheds going.

"Rhubarb is a gross feeder of nitrogen. Prior to planting we dig in wool shoddy from Bradford, and then we add chemical fertilizers as necessary. We use herbicides in the first year. If we can keep the field clean in the first year when the plants are young and lacking in vigour, then it's a lot better the second year, when the foliage is thicker. We also have to spray to stop aphids. If you don't bother, they inject the rhubarb with virus diseases and then you don't get as much growth.

"After the second year we start measuring the soil temperature from early October, using a thermometer that goes in four inches. We take a reading at nine o'clock every morning, and every degree below 49°F counts as one unit. We accumulate these cold units, and when we get to 200 we're ready to start breaking dormancy on the early variety, which is Timperley. Mid-season varieties like Prince Albert and Reed's Early Superb need 400 units, so they stay out in the field longer. Then the main-crop varieties we grow are Victoria and Stockbridge Arrow, whose little yellow leaf is shaped like an arrow. These need 500 units. So we want the frost early, so we can get these cold units as soon as possible.

"We lift as much of the root as we can, without losing any of the soil around it, using a lifting machine. They weigh about 4 stone each, on average. Then we bring them in the sheds with the tractor and stand them down on the floor by hand, as close as we can, leaving a main path down the middle and little paths off to the side for the pickers. Then we seal the doors and put the heat on. Timperley will grow at about 50°F but the others need 50° to 60°, day and night. We water them, but they don't take up any feed at all. They don't respond to treatment in the shed. All the goodness is in the root.

"It takes about three weeks on earlies, from the first heat to the first pull, and four weeks on the others. Then we pick over the roots once a week, about four times altogether. It takes a whole day to pick a shed and we want to keep the light off it, so we just have a very small bulb that hangs where the pickers are working, and moves along with them. Then we'll put a candle on the path for the person that's carrying it into the packing shed for grading."

Mr Oldroyd opened the tiny side-door of a shed of Stockbridge Arrow. He lit a candle and placed it in a purpose-built holder on the end of a yard-long metal stick. He stuck this in one of the roots and bent to pull a dozen sticks for me to take home. The shed extended a hundred and fifty feet from the small flickering light, and was filled with 15,000 roots. In the warm, humid darkness, with every sense stimulated by this strange vegetative growth, it was possible to contemplate simultaneously the brick and timber of Dickens and the living Triffids of John Wyndham. Out of the gnarled, orange-brown roots shot bright red sticks of 'fruit'. Some were just bursting from their crowns, some were two feet high; some stood bold as pit-props, others tentative as pipe-cleaners; and on top of each, the tiny, crinkled yellow leaf, like the overblown antenna of a more sensitive being, each facing a different direction. In the flickering light of the candle I thought I could sense them all twisting towards me, all alert to this brazen intruder.

"The roots are exhausted after about four pickings," Mr Oldroyd said after our hasty retreat. "So we take them out and put them on a heap to rot down and then if there's no disease in them, they're spread on the land as compost.

"We don't particularly aim for the Christmas market, because we think there's a lot of variety of fruit about, at Christmas. If we were going to fill the same shed twice, then we'd be looking to start pulling in early December, so we could start with another crop in January. By March they start with outdoor rhubarb in Kent and Cornwall, although I've known it to be ready by the back end of February. Then we can virtually close up."

Ken Oldroyd also grows outdoor rhubarb, which is ready by the end of March or early April, and a variety of outdoor vegetables, mostly Brassica. Much of this produce is marketed through Yorkshire Rhubarb Growers, a co-operative including a dozen other rhubarb growers: "This has brought a great deal of stability to the marketing of rhubarb. Rhubarb has become a bit of a luxury, price-wise, and that's why we have to keep our quality very high. We must get into these new varieties which are being bred at Stockbridge House by the Ministry — like Stockbridge Arrow. The old Victoria is becoming grown out, it's becoming pale as it's forced, but these new varieties are red, really red, and that's what the customer is looking for."

Mr Oldroyd went off in search of the trophy he won as World Champion Grower in 1968. There are prizes at the February Show of the Rhubarb Growers Association in a host of categories: six sticks of Prince Albert weighing over 5 lb.; six sticks of Victoria between $3\frac{1}{2}$ and 5 lb.; six sticks any variety . . . and the same few family names appear in the honours lists year after year, decade after decade. But it is hard work, even if you can do it with your coat off. Mr Oldroyd is not at all sure that his son and son-in-law, who work with him now, will continue with it after he retires.

There are also prizes for the best pie; the best cooked dish; the best cold sweet or drink. Mrs Beryl Oldroyd showed me pictures of her Union Jack cake, using stewed rhubarb thickened with gelatine for the red parts, which won in the year of the 'I'm Backing Britain' campaign. When Sir Francis Chichester sailed round the world single-handed, she won with a sponge *Gypsy Moth* in a sea of rhubarb.

Rhubarb does not rate highly as far as vitamin and mineral content goes, but it does have vitamins A and C, and a range of minerals in small amounts. And these nutrients become more significant when compared to those in bottled or canned fruits. English cooking is at its best in cold weather, and fresh rhubarb can provide a lovely sharp, refreshing flavour at the end of a hearty winter meal. Mr and Mrs Oldroyd's daughter, Linda, is a home economics teacher at the local high school, and she includes the versatility of rhubarb in her teaching. It is a good mixer with other fruits, especially oranges, and with cinnamon and ginger. Stewed rhubarb, heated very gently so that it stays in pieces, with a little demerara sugar, is also excellent with breakfast cereals. And although the days may have gone when anxious teenagers would consume vast quantities of rhubarb to "purify the blood" for the sake of their acne, still some children like to dip a raw rhubarb stick in sugar, to suck like a lollipop.

"Can rhubarb stick up for itself?" is the headline for one of the recent articles promoting the fruit in the trade press. Cooking apples emerge from their cold stores at half the price of fresh rhubarb, with its intensive labour and high fuel costs. "I don't think anyone could come into the rhubarb business now and show a big profit," says Norman Asquith. But the man from the Ministry is busy trying to engineer the shorter stem which the supermarkets prefer . . . still breeding for the future.

3

Cleansing the Blood — Watercress in Hampshire

THERE IS A rural tradition of cleaning and thinning the blood which long predated the urban treacle and brimstone. In the American Deep South there is a craving in spring for pokeweed to thin the winter's clogging, green-less lethargy of salt pork, fat back and dried beans. A Frenchman wrote that his mother used violet leaves for the same job. Sorrel, mustard greens and nettle broth all have their champions. There was a time when, according to one historian, prehistoric youngsters were so eager for spring greens that when their fathers brought home their kill, they would eat the partly-digested greens in the animal's stomach before turning to its flesh. But in Britain the candidate with the most illustrious past and promising future is pungent, deep-green, fresh watercress. There is even a kind of dogfood which has watercress in it.

And a piece of fiction: "The Watercress Girl". "'Oh Christ', she breathed," as the judge delivered her sentence for throwing acid in the face of her lover's betrothed, "for it was the lovely spring; lilac, laburnum, and her father wading the brooks in those boots drawn up to his thighs to rake the dark sprigs and comb out the green scum." There is no more passionate character in all of A. E. Coppard's stories, and he gave her watercress to cut for a living.

The watercress farms of Britain are strung out from Kent to Dorset on the edge of the chalk Downs because here are the fresh-water springs, and above all else watercress needs a vast amount of pure water. I visited the Hampshire watercress beds in early April. Having left the train at Andover junction, near Abbotts Ann I turned down Cattle Lane, a winding narrow road leading up the shallow Anna valley. A row of tidy houses with large vegetable and soft-fruit patches vying for priority with lawns and rosebeds gave way to smallholdings. I passed a flock of hens in a large run. Three young calves basked in the sun, lying on a pile of hay in one front yard. There was a nursery with a half-acre of broken glass . . . a sign in chalk on an old pallet, advertising potatoes and carrots for sale. . . . These smallholdings were relevant to the watercress, too. Ironically, large-field farming now means a more intensive use of chemicals, and modern watercress production involves constant vigilance against pollution. With smallholdings as neighbours it is much safer.

Now the lane ran beside the Anna and I saw my first cresses. They were escapees. It takes only a piece of stem to get into the river and latch on to the bank, and the cress will grow. It has been growing wild in other British streams since prehistoric times, and if water-cress had a role at the ancient celebrations of the Spring Equinox, it must have appeared and tasted to the Britons then much as it does to us now. Unlike most of our cultivated foods, which have been so improved botanically that they are hardly recognizable as the same species, our cultivated watercress is practically identical to its wild ancestor.

As I continued upstream, both banks were lined with cress. Thick floats of leaves reached almost to the middle. I was getting close. I was not tempted to reach through the hedge and pick some, however, for I had been reading the National Farmers Union/Watercress Branch *Code of Practice*, which makes eating wild watercress sound inadvisable.

I might have been excused for thinking that my doubts were well founded, as I turned a bend and saw through the hedgerow what appeared at first to be a sewage treatment plant . . . a rectangular bed of gravel, covering some third of an acre, with low concrete walls and two little windowless, unmanned kiosks . . . pumping stations . . . with dark green doors and mysterious numbers on plaques . . . like '37' . . . And as I progressed, another! Three, four, five more beds . . . stretching as far as I could see, side by side in perfect geometric order. Now I could hear a persistent trickle . . . water without obvious source, water put to work. . . . The next bed was covered with a translucent film of something greenish. . . .

A lorry came towards me, filling the lane, and turned at the last moment to cross the Anna on a narrow bridge and enter the plant. In huge letters down the side of the vehicle I read Hampshire Watercress and the trademark Vitacress.

A caravan parked against the back of the packhouse served as the works canteen, and I was shown inside to wait for the manager. I might have been in the centre of an industrial estate in the West Midlands.

The watercress men like to present themselves as a modern

industry. But folklore can be useful too. I remember a colourful wrapper from a batch of Sicilian lemons — the kind of paper which my mother, in her youth, would have threaded together to hang indispensably in the outhouse — which boasted:

It is gathered with the utmost care, so that none of its superb naturalness is lost consisting in: a very thin and uniformly yellow peel, with a great deal of juice and very few seeds. Its rich contents of carbohydrates and organic acids such as: Glucose, Fructose, Saccharose, Citric Acid, Malic Acid, and such vitamins as: Vitamin C, Provitamin A, Thiamin, Niacin, Inositol and many other beneficial substances help cure: cancer, cholesterol, quinsy, poisoning, botulism, bronchitis, kidney stones, lithiasis, cirrhosis, cardiovascular illnesses, hypertension, constipation, corns and calluses, diabetes, diphtheria, migraine and headache, fevers, feverish ailments, gingivitis, stomatitis, glossitis, chilblains, obesity, pneumonia, rheumatism, tuberculosis, and many other sicknesses. It is a healthy fruit essential for one's daily diet.

The watercress publicity machine at Agriculture House rivals this Latin extravagance, but with claims more in the British national character:

[Watercress] can make a very positive contribution to the dietary wellbeing of the nation. . . . It has a Vitamin C content that is impressive, an even bigger Vitamin A content, up to 10 times more calcium than a similar weight of any other vegetable, iron, riboflavin and dietary fibre. . . . That collection is reckoned to make you bright of eye, supple of skin, be good for bones and teeth, help ward off anaemia and generally keep you in the pink. . . . There's an added bonus, too. Watercress contains Vitamin E, reckoned essential for a good love life!

This is a good example of 'respectable' claims made on the basis of a laboratory report on nutritive values: a kind of instant folklore.

The Romans thought watercress helped them to make bold decisions, but mixed with vinegar they used it to treat mental illness. Perhaps this was the first official recognition of the 'mad dog = good general' equation. The Elizabethans variously promoted watercress potage for lethargy, scurvy and toothache. In the seventeenth century the cress was eaten to stop hiccups, and crushed leaves were applied as a compress to remove freckles. My grandmother can recall watercress tisane, or tea, being an effective cure of rheumatic pain. It was also taken for migraine. A stronger brew (½ pint of boiling water poured over one bunch of chopped watercress, left to soak for ten minutes, cooled and bottled) was used to sponge aching brows and temples in rural homes at least until the 1940s.

Ken Marsden smiled rather mockingly at the idea. He conceded that watercress sales are "riding high on the health food fad" but he sees that more in terms of slim young women watching their calories than grandmas sipping herbal teas. He represents the 'modern industry' viewpoint.

"But what about your staff?" I asked, thinking of the garlic growers who swore by their product: "Never a cold all winter." "It saw me right across Africa, without as much as diarrhoea."

"They see it as a job, that's all. It's just a job."

Mr Marsden is a packhouse manager. The company is owned by Malcolm Isaac, who started with 1½ acres of cress in 1950. Now he operates 65 acres of beds at nine farms and has become the largest watercress producer in the EEC.

The variety is called American Dark Green but as selections have been made on the Vitacress farms for many years it could now almost be called the Vitacress strain. A propagation department supplies plants to all the farms in Hampshire. The cycle varies according to season from six weeks in the best growing time to as many as fifteen weeks in the dead of winter. In the glass house seed is sown regularly on either peat or a polymer gel on plastic trays. These are moved down the glass house as new trays are planted to emerge the other end as 2-inch seedlings some ten days later. The seedlings are transferred to nursery beds to acclimatize them to aquatic conditions. When they have hardened off and established a root system they are transferred to the commercial cress beds.

At Abbotts Ann the water is drawn up by 20 pumps. During winter the beds are flooded and the cress grows under the slowly moving water. It is pulled up at harvest point, and the roots trimmed. In spring the cress grows up out of the continuously flowing water and is cut from the bed. At this point they were using about a quarter of a million gallons per acre. The water is free, but can be used only under licence from the River Authority. Every week the company chemist checks for pollutants and scrutinizes mineral levels in case trace elements are lacking and need to be added. The cress could not be described as 'organically grown', but I was surprised by the relatively low level of chemical manipulation for such a modern intensive crop. Sometimes, when the cress is growing fast, superphosphate is broadcast in granular form. The cress is prone to crook root, which is the accursed club root familiar to growers of all types of Brassica. Then a zinc drip is added to the water. Watercress is also attacked by greenfly and aphids, which cause the leaf to curl and stubbornly refuse to wash out. Pyrethrums are used at low volume to control pest populations when necessary but there is no blanket spraying.

Developing disease-resistant strains is the esoteric end of things. But there is also the daily management of the beds . . . the efficient use of space and labour. . . . Twelve years ago the yield was 35 bunches per square yard and now it is 45. The target is 60. "We have some beds permanently covered with polythene," Mr Marsden said, "but the other approach is through Portugal." Hampshire Watercress Ltd now crops 6 tons of watercress per week through the coldest winter months in southern Portugal. This is airlifted to Britain for the supermarkets. A prime objective in the supermarket trade is continuity of supply, and this is the great achievement of Hampshire Watercress. The retail price is more-or-less constant throughout the year. If they can sell an extra 6 tons per week in January and February, then they can hope to achieve a peak 6 tons per week higher in the spring and summer.

At the end of the line a bed had been cleared and the plants pulled out. A 'production team' of three young lads in wellies were topping up the gravel from the back of a tractor trailer. Then it would be raked and rolled and replanted. Traditionally about five crops are cut from each bed before it is stripped and replanted, though it is possible for a good bed to give ten crops a year, and to go on yielding for ten years.

On the majority of watercress farms the cress is cut by knife in small swathes called 'hands' and laid carefully in layers in bins so that it can easily be tied into bunches prior to washing. These traditional bunches are sold on the wholesale markets and still account for about 70 per cent of total sales. But supermarkets consider these bunches inconvenient, as they can occasionally drip water and become messy. So polythene pouch-packs, or stretch-wrapped pvc rigid trays, have been developed, and since the cress is loose and can be of different lengths, machine harvesting becomes feasible. At Abbotts Ann they use a machine specially developed by Hampshire Watercress. It consists of a fixed-blade grass-cutter mounted on the front of a small tractor, beneath large suction ducts which run along each side of the tippable containers at the rear, so that the cut cress is 'vacuum-cleaned' from the bed.

We moved on to the packhouse where I was issued with a gown and cap. The rules were strict enough outside! Cultivation of watercress under 'hygienic' conditions began in Germany about four hundred years ago, with the first commercial beds in Britain constructed in

1808 by nurseryman William Bradbury at Springhead, near North-
fleet in Kent. Beds were typically surrounded by earth dykes as
growers dug out pools next to streams. Up until then watercress had
been gathered from wild beds in streams, and was often sold with a
fair percentage of kingcup and water-marigold leaves. Now the beds
must be of gravel with permanent impermeable sides, usually of
concrete, and they must be fenced against animals; birds must be
excluded and vermin destroyed; surface water must be intercepted and
diverted; beds must be clear of adjacent muddy areas and not liable
to flooding. These rules of the 45 grower-members of the NFU
watercress branch were largely devised to eliminate the threat of
liver-fluke, a parasite carried from infested animals via a tiny snail.

Mr Marsden admitted that the last case of liver-fluke contamination
in watercress was in 1935, and that no wild watercress, which might
possibly be contaminated, now appeared on the market.

Once inside the sheds the strict routines were maintained on the
grounds that watercress is usually eaten without cooking, although if
these immaculate conditions were mandatory for all fruit- and salad-
packing houses I suspect there would be hearty dissension within the
NFU.

"These machines process 120 packs per minute, 15 hours per day.
We work an evening shift from March to October. . . . The cress is
vacuum cooled if not packed immediately. . . . Watercress is one of
the most perishable crops any retailer sells. If it's sold through Marks
and Spencer, it's under refrigeration from twenty minutes after it
leaves the bed right through to the point of sale. "We can't cut down
on the twenty minutes," Mr Marsden added. "The women wouldn't
stand working in the fridge."

The woman at the start of the line was loading loose cress into the
wash tanks.

"Is it cold on your hands?" I asked.

"On a cold morning," she answered.

But Mr Marsden put me right: "The water emerges from the wells
at a constant ideal temperature of 51°F, fifty-two weeks a year. . . ."

The watercress passes through a second tank of chlorinated water
chilled to just above freezing-point, and then to an upper level where
the water is shaken off by vibration. Women feed the cress, about 3 oz.
at a time, into cups on a revolving turret. An automatic plunger forces
the cress down into a sheet of polypropylene film which is then tied
into a pack, weighed, and labelled, all by machine. I pondered briefly
which new techniques or improved efficiencies Mr Isaac could bring
back with him when he visited the only watercress growers in the

world who were bigger than himself. "The boss has been to Florida twice," said Mr Marsden. But it transpired that the industry in America was far behind that in Britain, and if anything the American growers were picking Mr Isaac's brains.

Keith does the Liverpool run.

According to Mayhew's *London Labour*, written in 1846, "the first coster cry . . . heard of a morning in the London streets is that of fresh 'wo-orter creases'".

Keith drives overnight, dropping his consignments at the super-market distribution depots at all points towards Liverpool. He has just arrived back mid-afternoon, having kipped in his sleeping-bag on the Leyland's cot when he ran out of hours. He is married, with young children, like most of the drivers. "It's just a job," he told me, unprompted. "You soon get used to nights. When you're working away three nights, it's amazing how quickly a week goes." Now he's taking the London cress to another farm, to trans-ship to the chap who does the London run, before he knocks off.

We British eat 4,000 tons of watercress each year. That's £8 millions' worth. But it works out at less than one bunch per person. That is almost certainly going to increase, whatever the efforts at Abbotts Ann, as our increasing consumption of salads continues. We may not put it in our soups and soufflés, flans and omelettes, or dressings and cocktails, as the Watercress Cookbook recommends, but it is going to find its way into more salad bowls and sandwiches. And an excellent thing, too.

The company operates thirteen refrigerated vehicles and distributes its cress all over England. No more watercress in the guard's van on British Rail. And no more control of the market from Covent Garden. For Scotland they trans-ship to contracted hauliers.

As I drove with Keith on the main road in his 16-tonner truck, I asked him how he liked the Leyland. "It's more powerful than the two new DAFs," he answered. But then he went on to extol the hi-tech advantages of the rival machines in the fleet: "They've even got heated side-mirrors. . . ." I asked him how it handled on the motorway in a strong wind, thinking of the light load and high sides. But Keith corrected me. A part of the load was always the old-fashioned bunches, for the wholesale market. They're heavy, be-cause they sit 6 lb. of crushed ice in each box, to melt gradually and keep the cress fresh in a trickle of cold water. Just like the pre-war California lettuce, I thought, earning its name Iceberg in railroad boxcars heading for the big cities.

4

More Good Tips — Asparagus in East Anglia

I ARRIVED AT March on the edge of the Cambridgeshire fen, travelling again by train, and talked to an old railway worker who was sitting on one of the platform benches. He told me of the days when March, at the convergence of five railways, had more sidings than any other marshalling yard in the country. There were a hundred 'rows', each with two 'humps' for uncoupling, and they would handle a thousand wagons each shift. Much of the traffic was bringing coal into East Anglia, and taking agricultural produce out. "Now there's only ten rows for arrivals and ten for departures, and it's all air-brakes and no mucking about. Two hours and the train is out. But in the old days there'd be wagons chock-a-block over a hundred rows and it might take a week for a particular wagon to work its way through." I wondered what a truck-load of asparagus would look like by the time it reached Manchester.

Asparagus, however, is a relatively new crop to the area, and in any case would always travel as 'passenger parcel'. Mrs Aveling gave me a ride from the busy centre of the market town a mile or so eastwards to Badgeney Farm. Her husband was supervising last-minute preparations for an open evening held by the Agricultural Development and Advisory Service, in conjunction with the Asparagus Growers Association. The programme included lectures, discussion, and an inspection of the variety trials currently underway at the farm. There was a frantic rush to get the work done and the packing shed ready. The bulk of the morning's picking was in the cooler and the boxes, knives and scales were being packed away. 'No smoking' signs were going up, a screen and slide projector brought out, chairs set in rows. An evening shift would come on later, when the packing-line was restored. For asparagus is a crop with its own imperatives and during its brief harvest period it dominates life on the farm. I watched as the women sorted the last of the morning's spears into different grades, trimmed off the stalk to a uniform length, and then arranged them into neat bundles and carefully tied them round like sticks of dynamite . . . the delicate tips, so easily broken, seemed primed, like fuses . . . and the women passed the finished bundles along with both hands, packing them like rounds of live ammunition.

Outside the shed I was startled by a low-flying jet belonging to the United States Air Force which came as though from nowhere, loomed suddenly overhead so close that I felt I could reach up and touch it, and then shot immediately and totally from view leaving only a deafening roar as proof that the event had existed.

I walked out on a sandy track between sunken ditches, past fields of strawberry plants and corn, until I came to the asparagus. Now the enormous, clear sky which has drawn artists to East Anglia for centuries began to exert its awesome influence . . . and under it, like point and counterpoint, the profound strength and fertility of the soil. Asparagus spears thrust their way from one to the other. Here and there, thin wisps of 'sparrow grass' or 'sprue' had already grown to fern, but down the main lines of the rows, thick shoots stuck out like pokers . . . or fat green candles with a tinge of purple flame . . . primitive symbols of sheer rude strength and vigour. Asparagus is the food most frequently cited in accounts of the medieval Doctrine of Signatures and it is easy to see why it was assumed to possess specific sexual powers, and why it has been valued through the ages as an aphrodisiac.

Asparagus is a genus of the lily family (Liliaceae) containing about 150 different species which grow wild from Siberia to South Africa. Several African species are grown as ornamentals but there is one species only, *Asparagus officinalis*, which is grown as a vegetable for its succulent young shoots. This plant is found especially in sandy soils around the mouths of rivers, on Mediterranean coasts and the Atlantic shores of Europe, including those of Britain.

There are paintings of bundles of asparagus, trimmed and tied just as it is today, in Egyptian tombs dating from 3000 BC. The ancient Greeks gave us the word 'asparagus' but they used it to mean any tender vegetable shoot which has not yet broken into leaf. The Greeks certainly used the specific plant we now call asparagus as a medicinal herb, but the earliest written comments on its use as a food come from Roman writers. Cato described a method of asparagus cultivation which remained the model until the nineteenth century; Pliny recorded that market gardeners on the sandy marshes near Ravenna — which still produces the best asparagus in Italy — were growing spears so fat that they weighed three to the pound; Julius Caesar expressed a preference for asparagus to be served with melted butter; and Caesar Augustus, presumably in revolt against limp, overcooked spears, is credited with the catchphrase, *'velocius quam asparagi coquantur'* — 'faster than you can cook asparagus'.

After the fall of Rome, widespread cultivation of asparagus, calling for well-attended permanent beds, fell into decline throughout Europe, although wild plants and naturalized escapes from previous times continued to be used medicinally through the Dark Ages. The crop was re-introduced as a food via the Moors in Spain, who brought roots from the eastern and southern Mediterranean. By the seventeenth century it was once more an established luxury in many gardens, so that Samuel Pepys could record, in the late April of 1667, that he bought a hundred sticks of asparagus at Fenchurch and 'had them with salmon'.

By the beginning of the nineteenth century more asparagus was grown in Britain than in any other country. Small pockets of expertise developed where the crop was concentrated: growers in Essex and Kent supplied the London market; the sandy soils around Formby, near Southport, fed the North; and the heavy clays of Evesham inexplicably produced nearly half the national tonnage. Marjorie Hessel Tiltman in *English Earth* described what was possibly the first example of the modern 'farm-gate' phenomenon, in the 1930s:

> Evesham has an asparagus show, an asparagus day, an asparagus week. Most people grow asparagus. Many people sell asparagus. Every other garden bordering on the main roads has outside it the little wooden stall which the motorist has rendered so profitable a business venture. Sometimes a shy child waits there, with a little wooden bowl beside her for the takings. Sometimes you must walk up the brick path . . . you may name any quantity and walk up with the grower to the bed itself and see cut, by expert hand, a bunch of the choicest, crispest stems.

Estate agents still list 'established asparagus bed' alongside 'all mod cons' for their truly desirable residences. And at The Round of Gras pub on the A44 just southeast of Evesham, landlord Buster Mustoe still has a high reputation for his luncheons and dinners during the season. Each dinner is served with three-quarters of a pound of asparagus, and there is an excellent soup made from stock and low-grade 'sprue'. In the six-week season Buster serves over a ton of fresh asparagus. He has also collected an interesting selection of memorabilia, including special asparagus serving dishes, and a photograph of one of the last lorry loads — almost 8 tons of 'rounds of gras' — which was sent from the Vale to London in 1939.

At the start of the Second World War the government declared asparagus an unjustifiable luxury, and the beds were ploughed up in order to grow more urgently needed food. There was a poor market for asparagus long after the war, and smallholders baulked at the high capital costs involved in replanting, the great demands on labour, and the long delay before they could expect good returns. So the crop never regained its former glory in the Vale.

In the 1920s one Captain Kidner, who was interested in asparagus plant-breeding for improved varieties, brought the crop to East Anglia at the suggestion of the Ministry of Agriculture. And in 1933 Lord Fisher of Kilverstone made a considerable planting on the sandy soils around Thetford. It was in this region, and as a large-scale speciality crop grown by only a few farmers, that the more recent revival of asparagus eventually took place. About 1,200 acres of asparagus are now grown in Britain, with the vast majority in East Anglia, and fewer than 100 acres in Evesham.

In 1978 a young farmer named Michael Paske from Huntingdon won a Nuffield Farming Scholarship to study asparagus production in the United States and Mexico. With single fields measuring hundreds of acres, and a total of 50,000 acres in California alone, he found that the American industry had developed since the war with the help of a growers' association which co-ordinated research into cultivation and marketing. Soon after his return, Michael founded the UK Asparagus Growers Association which now has 81 members representing three-quarters of British production. They pay membership costs of £15 per year for the first two acres, then £5 per acre.

Michael's father had travelled abroad for a paper and packaging company, simultaneously indulging his interest in asparagus-breeding by collecting various varieties. One of these he hybridized with a variety of Captain Kidner's to produce Regal Pedigree, so named because it was first grown at the former home of Edward VII's mistress, Lillie Langtry. Michael grows Regal on some 60 acres near Huntingdon, and also on land in Jersey, in Yorkshire's Vale of Pickering, and Scotland's East Lothian. He has sold the variety to growers in Portugal, Spain, Malta and Scotland.

"Asparagus is a maritime plant originally, and Captain Kidner proved it grew very well on the sandy soils at Lakenheath. But we are finding now that anywhere you can grow good crops, you can grow asparagus. The main thing is to have good drainage. In Evesham they had to ridge it up to keep it out of the water, but here we can grow it in flat beds. It is also prone to frost damage at the start of the season, so you've got to make sure you're not in a frost pocket.

"Seed is sown in a seed-bed in the spring and the crowns — the octopus-like roots — are lifted the following February or March. These 'one-year crowns' are then planted out in blocks in the field where they might stay anywhere from ten to twenty years. The roots have to be spread out very carefully to avoid damage, and then covered to a depth of about five inches. By the following season the two-year-old crowns can stand a light harvest for a week or so, but we don't get a proper crop until the third year. So it's a long-term investment. It can cost £2,500 per acre to establish an asparagus crop.

"Then we get 30 or 40 spears over the season from each crown. They send up all types, including scrawny ones that have no value. Harvesting usually starts in late April and is very labour intensive. In warm, humid weather new spears can grow 9 inches in twenty-four hours and the tips begin to flower open, so the fields must be cut every day, including weekends and bank holidays. Then on cool days the cutters can find themselves walking miles up and down the rows with not much to show for it. They use all sorts of knives, although there are specialist blades with V notches at the bottom for cutting the stem just under the soil. Cutting after the end of June reduces the crop for the next year, and the quality deteriorates, so the spears which appear from then on are allowed to grow. That's when we put on nitrogen to feed the plants. They produce a beautiful fern seven or eight feet high. The roots bring up minerals and store all the nutrients needed for the spears for the following year. We scythe down the fern in January, after the frosts have wilted it, and get rid of all the debris. That removes the hibernation sites of the asparagus beetle."

Michael Paske sells his asparagus in traditional rounds; in vacuum-cooled, shrink-wrapped tray-packs for the supermarkets; and in 1-kilo bundles by mail order — by first-class post — all over Britain. He also markets the asparagus of several associates, including that from the 30 acres at Victor Aveling's farm near March.

Victor Aveling introduced the open evening at his farm, and welcomed the 60 or 70 asparagus growers and their wives who had travelled from as far afield as Hampshire, Somerset and Gloucestershire.

"I'm surprised that so many of you could come. With this warm flush we've had, I haven't seen so much asparagus for years. And next week we'll be swept into the strawberry season. So it's very hectic and if the farm looks a bit untidy, it's because of the rush. . . ." He gave a brief summary of his own set-up as an asparagus grower. "We've got plenty of labour across the train track in town, and they can walk or bike out so we've got no transport worries. . . . We've been growing asparagus here for nine years and we've got 30 acres of it. Then they started dying. The Cambridge man helped and ADAS has been very helpful. And whereas we had pockets of expertise before, now with the Asparagus Growers Association much more information is available . . . When the University of California man saw it, he said if you've got Fusarium, you should plough in. . . ."

A significant part of the programme which followed was a report on Fusarium wilt from Dr D. R. Ellerton of the Plant Pathology Department of ADAS at Cambridge. He told us that most growers have got this major disease of asparagus in some shape or form, although it had not been considered prevalent in Britain until recently, and growers had attributed plants dying-off to other causes. The fungus responsible spreads from the mother plant through the seeds and cannot always be seen, but will quickly become evident when the plant is under stress. Red streaks appear at the bases of the shoots and the inner root tissues collapse. He described the chemical treatments of seed which would restrict contamination from this source, and stressed the importance of planting on land virgin to asparagus. He then reported the results to date of the various trials of chemical treatment conducted on one-year Regal crowns at transplanting, and on an infected established crop at Victor Aveling's. Sheets of data were gratefully received by those present, showing, for example, that a Tecto acid dip and drench increased yields for the transplanted crop by 115 per cent. Other papers described the potential ravages and possible actions against the asparagus beetle, asparagus fly, asparagus rust, purple spot, violet root rot, and slugs (several species). Reports were also given by ADAS officers of controlled trials of different varieties at Badgeney Farm, and elsewhere, prior to actually inspecting the trial site. There has been no breeding of asparagus in Britain since 1950, as the greatly reduced market did not appear to justify it. But in Holland, Germany, France and the United States there has been a great deal. Some of these programmes have concentrated on all-male varieties. In a traditional asparagus field, half of the plants will be female, producing berries and seed towards the end of the summer: in all-male plantations more of the energy of the plant is concentrated on producing fern, which subsequently feeds the root better for next year's crop of spears. All have aimed for higher yields of class 1 spears starting earlier in the year and cropping steadily, rather than giving flushes in hot, moist weather.

Fifteen varieties were planted as one-year-old crowns in 1981 — twelve inches apart and four inches deep, in rows a yard apart. The spears produced are cut, size-graded and weighed, and various comparisons of yield and costing made. The results cannot be properly evaluated yet, as the German all-male variety Lucullus, for example, is only fairly ranked at present, but is supposed to give increased yields up to the eleventh year. And Regal was the only variety to show a positive cash-flow figure by the third year, but this was because of the much higher price of the imported crowns, which will gradually cease to be such a heavy weighting factor in years to come. It seemed clear, however, that the new foreign hybrids are giving higher yields than the open-pollinated varieties, Argenteuil, Connovers Colossal, Regal and Saxon, which are the most common grown in Britain. And with the exception of Regal, the British gave a particularly small percentage of class 1 spears.

Another set of trials at Luddington, near Evesham, ranked the old local favourite, Giant Mammoth, lowest of all. But it showed that yields generally were higher in traditionally lower plant populations in Evesham than those at March, where spacings had been arranged to suit the control of weeds by mechanical cultivators which Victor Aveling used for his strawberries. A good six-year-old crown has over three thousand yards of root, which explained why greater spacing produced better results. At Luddington there is yet another trial, begun in 1983, of various hybrids created at the John Innes Institute. The Asparagus Growers Association discovered almost by accident that the Institute had chosen asparagus as the medium for a project on the gender of plants and had created 35 all-male lines. There is great potential here, though if the trials produce good results it will take three or four years to produce sufficient quantity of commercial seed.

There were many questions from the floor, all of them efficiently answered. . . . There was no need for irrigation, for example, after crowns are established. The spears are not transpiring, the roots are deep; indeed irrigation lowers the soil temperature and reduces growth. The mood was one of enthusiasm and even excitement. Relatively young growers, new to the crop, seemed eager to invest and expand, and were impatient that the precise advantages of this new generation of stocks were not yet conclusive enough for them to make immediate decisions. Mira seemed best for producing jumbo sticks. . . . Lucullus yielded particularly well in the sizes preferred by supermarkets. . . . Cito was the earliest to harvest. . . .

And then a grower asked if there had been any trials of consumer preference. A titter of amusement ran through the audience, as if this was a silly question. But there was a hint of embarrassment in it too, rather as though if this aspect *had* to be raised, it should perhaps have been done so earlier.

Impromptu taste tests for toughness and flavour had been carried out, we learned, and Lucullus, perhaps the hottest tip in growers' circles, scored poorest. "The Camden Food Research Association have all the varieties for professional testing," said the man from the Ministry. "We are interested and we are taking note of it."

In the opinion of Michael Paske, soil is more important in determining flavour than the variety. But there are undoubtedly characteristics which influence consumer preference. Regal, for example, is all green and tender while Argenteuil has a mauvish head. And with Connovers Colossal the bud is more open, which some dislike and others favour. In most of Europe asparagus is entirely white, being blanched by earthing up to achieve greater tenderness and a subtler flavour, at least in the opinion of its advocates.

For some of us, the arrival of British asparagus on the market is a great portent of summer, like Jersey Royals, and a time for a treat. But with an average consumption well under one spear per person per year, even with considerable expansion, fresh asparagus will remain a luxury crop. And since in other producer countries, such as the USA, Mexico and Taiwan, the period of photosynthesis is longer and the dormancy period shorter, their greater yields are likely to give them a continued edge on the market for processed asparagus. There are those, of course, who will say good luck to them. These are the people who think of asparagus, like globe artichokes, as a mere excuse for consuming vast quantities of melted butter. Or, in the extreme, like P. G. Wodehouse: "Have you ever seen Spode eat asparagus? No? Revolting. It alters one's whole conception of Man as Nature's last word."

5

A Garden of Cucumbers? — Cucumbers on Humberside

―――――――――――――――――

CHRIS BEAN PUSHED open the door of his home with his elbow and made his way cautiously to the kitchen. Instead of proffering his hand for me to shake, he held both hands out palms uppermost, for me to inspect. They were covered with black stains. His blue jeans and plaid shirt, too, were streaked with patches of oil and grease. He stood still for several moments, at first a martyr on the cross, then more like a scarecrow of the machine-tool shop. At last he said with an ambiguous smile: "Behold! The modern grower!"

Chris is in his mid-thirties, and this was the first time I had seen him since we shared a memorable mathematics teacher some twenty years ago. He has an effervescent energy and he launched at once into an eclectic account of the growing of cucumbers. I remembered him at school drawing plans for a revolutionary hemispherical greenhouse which would minimize losses from reflection, and he laughed now as he told me that too much light is actually an embarrassment for the cucumber grower. He left Manchester University after one term, to join his father in the glasshouses. Since then, he has set up and then left a microcomputer business; launched and folded another company selling uninterrupted power supplies and regulators for computers; and dabbled in numerous ideas for energy conservation in modern glasshouses.

In the meantime there are cucumbers . . . about 2,500 boxes of them in a good week, from his 2 acres of glass. Or over 2 million boxes a year from the Humber Growers Marketing Organization which was formed by Chris's father and four uncles to market the produce of the cluster of glasshouses around Brough, near Hull, owned by members of the Bean family. But behind the effervescence, in Chris at least, there is a kind of malaise.

"In this business, if you're small you can't afford to make a mistake. Every year you invest God-knows-how-many thousands of pounds in oil, labour, seed and so forth before you see a return on your money. You've only got to have a knock, either a failure or a disease or something like that, and you're in it up to your eyeballs. A typical example is my cousin who pioneered rockwool in the group. He went in this year with all the confidence of three years growing on it, got a virus, and his yield dropped 50 per cent. Then you're working for the bank, you're just trying to get your money back."

He finally got his hands clean. "Today I was repairing a broken hose pipe. It's the devil to get a bit of pipe that goes in the hose to join it up. I had to go into town. Then my foreman rang me up and when I said I was repairing this damn pipe which has been wasting spray for the last month, he said, 'Well it's only seven feet from the end, I was going to cut it off.' I could have wrung his neck! He could have done that a month ago! But you get this kind of thing when you're under a lot of pressure. You've got to keep on top of the harvesting all the time. All the grading systems get overloaded."

Besides his foreman, there are five other staff, mostly women, involved with picking and packing. Chris doesn't do any of this work himself.

> I've got to keep my head clear, so I can keep an overview, so I can see disease and things like that . . . and machines breaking down . . . In the old days they used to reckon the money one-third oil, one-third labour and one-third profit. My greenhouses were built in 1966 for instance, when the oil price was 3 old pence a gallon sort of thing. That's all gone, but it was on those cost calculations that greenhouses were designed. We'll be putting heat on at night now. We track the outside atmosphere to minimize the fuel. And I'm going to install underground heating, which is a great advantage at the back end when the nights start getting cooler. But no sooner have you found a way of economizing on your fuel than the oil boys zip the price up.

★

The cucumber earned a place, surprisingly, on the short-list of goodies yearned after by the Israelites during their forty years in the wilderness: "We remember the fish, which we did eat in Egypt freely; the cucumbers and the melons, and the leeks, and the onions, and the garlic: But now our soul is dried away." And yet when the prophet Isaiah bemoaned the sinful nation which Israel had become by 740 BC, he described the remnant which survived amidst the desolation: "And the daughter of Zion is left as a cottage in a vineyard, as a lodge in a garden of cucumbers. . . ."

In fact, these ancient cucumbers may have been a particular type of melon. According to the plant historian Edward Hyams, the species *Cucumis sativus* to which our numerous varieties of cucumbers belong, is not native anywhere west of India, and probably never has been. The species *Cucumis melo*, or melon, on the other hand, which originated in Africa, includes a variety which is still known as the Egyptian Cucumber. It is very difficult to distinguish remains of different members of the large family Cucurbitaceae which includes both these species, and artefacts found at archaeological sites in Egypt may just as well have come from melons, or pumpkins, as from cucumbers.

It seems likely that our cucumber originated in Burma, or thereabouts. It was probably first cultivated by the people who built the cities of the Indus Valley, and only gradually was passed on to the west via trade connections.

Although the ancient Greeks and Romans may have grown cucumbers very similar to our own, the plant was not introduced to Britain until the sixteenth century. For almost three hundred years it was viewed as a great delicacy and was eaten only by the rich. Then a combination of glasshouse development, breeders' interest and Victorian enthusiasm created a popularity in which the cucumber became extremely fashionable. White, yellow, bronze and even bluish-skinned varieties appeared on the market. Favourites like Lord Roberts, Dr Livingstone, and Long Gun, as well as the Telegraph and Butcher's Disease Resister, still listed in some seed merchants' catalogues, became household names. Dainty cucumber sandwiches may have been, in Oscar Wilde's words, "a reckless extravagance" when Algernon ordered them specially for his visiting Aunt Augusta. But he wolfed them down with familiarity.

The Bean family started growing cucumbers on Humberside in 1961. "My brother liked growing them and there was plenty of profit in them, it was as simple as that," explained Chris's uncle, Johnny Bean. "We were peasants, living over the road there with five or six cows, delivering milk round the village. We had an acre of potatoes. We grew cauliflower, lettuce, radish, leeks, all that sort of thing. Grandfather took the stuff into Hull and he'd fall asleep on the way back, but the horse knew the way home. We always made our money in a bad year, when other people didn't irrigate, didn't do the job properly. Sometimes you were making big profits and you were wondering how to get your expenses up. Then it turned full circle and you had a job to make it pay. Then in 1932 we got a Dutchman over from Holland, Peter Los, and started growing tomatoes and all sorts

in Dutch-type greenhouses. They were made out of sides of railway
carriages sawn up to make posts and that sort of thing . . . cheap . . .
and it grew from that. There wasn't a greenhouse here until the Dutch
came. After 1932 Dutchmen got off the boat from Holland and looked
for a bit of land and that was it. They sold their stuff to the boats and
liked the look of the pennies. A lot of those families are still around,
still in business. Geest came selling wheelbarrows.

"Glasshouse men tend not to like field farming and vice versa. My
brother, Robert, was the first one who packed up all outside veg and
did nothing but greenhouses. It took another five years for the rest of
us to do the same, and grow nothing whatsoever except greenhouses." He corrected himself to say "stuff in greenhouses", but I
wondered if the first statement wasn't as accurate as the second: "For
eight years we had money on deposit. We'd run out of tax allowances
so we thought, well, we'll build some more greenhouses. Then we
grasped the opportunity when grants became available to get on and
expand as fast as we possibly could." Now there are 130 acres of glass
in the area committed to Humber Growers.

Chris took me on a tour of some of this glass, including his uncle's
25-acre complex, the Crystal Heart propagation business run by his
cousin Philip, and a 6-acre greenhouse which was formerly part of his
brother's holding. But we started with his own modest 2 acres,
situated close to the simple old farmhouse at Poole Bank where the
"peasant" Beans had been born.

"Planting varies, but I'll run through a typical season when my
father was here at Poole Bank. He grew on straw bales, which is a
cultivation whereby you steam-sterilize the soil and then dig out
trenches and put in straw wads, or bales. You put some fertilizer down
and water it in, then cover with soil and plant the cucumbers on top.
They root through into the bales. The straw heats up as reactions take
place and you've got effective artificial underground heating for the
roots and they just love it. They run wild on that. Nutrients come
from the chemicals and the rotting bale. In a good week you might get
something in the region of 1,000 boxes an acre. On rockwool in a
good week you'll get 1,500. There's an enormous difference in yield,
but you've got to have sophisticated equipment to do that. The
rockwool is inert, it's just processed rock to hold the roots. You put
the feed in the water. Then there's NFT, which is the nutrient film
technique, where you have a channel constructed of black polythene
on an incline and you put water with the feed in at one end and take it
out at the other and recycle it, and just stick the plants on the

polythene. You can grow tomatoes well like that, without wasting any fluid, but it's difficult with cucumbers because I think the measurements are more critical, and if you get a disease problem it spreads like crazy. Using rockwool, the medium itself is expensive, as well as the feed, but you can sterilize it and it can last you three or four years."

The greenhouse was dense with foliage, some of it yellow, as the plants were already past their prime. Even so, I counted nine long, perfectly straight cucumbers, ready for picking, on the first plant. At a lower level there was a network of wide pipes, for heating, and narrower tubes bearing fluid full of chemicals, with an open nozzle poised an inch or two above the base of each plant. The rockwool, a light grey, granular substance not unlike roofing insulation in appearance, was laid on top of polythene in shallow trenches about a foot wide. To isolate the plants from diseases in the soil, the entire ground surface was covered with polythene, costing £500 per acre but nevertheless boasting several yawning tears by this stage of the season. Concreting is not a more permanent solution because various undesirable organisms apparently thrive on it.

"I've tried rockwool this year and we had some good yields early on but we came in too late and we've ended too early. Normally we'd be sowing in December, although it's more economical for me to buy plants from my brother David's prop. house. We'd plant around February and should start harvesting six weeks later. Then we pick the same plants all summer. Once they've grown up to the support wires we tie them, and stop them, and then we might occasionally go down and prune. But someone who knows what they're doing with the feed can actually control the growth." We moved to a more sophisticated nursery nearby. He showed me the tanks of calcium nitrate and phosphoric acid, and the computerized control, which would time the feed: so many minutes for each plant. He read the pH of the feed from the printout, and the temperature, humidity, wind speed and wind direction.

"All you are doing with a computer is making the job easier. It's a matter of how people appreciate control."

I asked him about control of pests and diseases and he took me over to the eight small greenhouses operated by a Humber Growers entomologist and paid for by members of the organization on an acreage basis. Here, red spider mite predators and white fly parasites are bred on runner-bean and tobacco plants, to be released into nurseries when required. "This idea has been around for a long time, but it wasn't taken seriously until we had a very bad attack of white fly, brought in by Uncle Willy on some pot plants." This form of biological control is much cheaper than using chemical sprays, and the unit now sells predators all over the country. I asked Chris if he didn't also find attractive the ideology behind using a harmless, natural predator.

"It's money," he answered with a hearty laugh. "It all comes down to money. There's no sense of fair play in this business. If it makes some money, we'll do it." I mentioned that I was going on to visit an organic grower the next day. "Well, the plant isn't interested whether its molecule of potassium comes from pig manure or a sack of potassium nitrate. That's the thing about a plant, it's apolitical. It doesn't care. You're looking at a different philosophy of growing, a different way of getting that molecule to the plant at the right time in the right quantity, and the right ratio between it and say nitrogen and so forth. There are certain combinations which will not allow certain molecules to be released to the plant. For example, if your pH gets too high and you've got a particular ion in the root medium, it will lock it out. It won't take it up. Rockwool gives the plant everything it wants at the right time. You've got chemists on the job and they can work out what kind of environment you need at the root and you can *control* it as opposed to putting manure on, which is an art. You can have a healthier plant on rockwool because you can put in the elements in the right proportions so that it can actually fight off disease." Then he laughed again, at himself. "You *can* have very healthy plants, although if you look at mine just at the moment you might not think so, because I haven't fed them for the last two or three days. . . .

"All I would say to your organic grower is, how many tons are you getting an acre? I would imagine we're getting half as many again on rockwool. If his fruit is better, he'll have to command a price 50 per cent higher. Then we come down to the customer. The customer doesn't know it's organic or what."

Then Chris took me down to his Uncle Robert's 25-acre nursery, where £500,000 has been spent on three British-built, coal-fired boilers. "You can get grant-aid, if you've got good balance sheets," Chris said enviously. "But even then it's one hell of a capital outlay. The amount of hardware you need to create a certain amount of steam is bigger if you're using coal. And coal is dirtier. But you don't mind getting dirty if it's going to save you some money, and coal is at least 25 per cent less expensive, per therm, than oil."

To keep the boilers fired, they have a stockpile of about 400 tons of coal and, at the peak, three trucks carrying 20 tons each will pull into the Carrdale yard each day. But the coal is lifted into hoppers mechanically and then falls through sealed tubes — I could see no evidence of dirt myself, and with modern scrubbers and 82–86 per cent efficiency, they don't anticipate any environmental trouble. Carrdale, in fact, seemed to me the epitome of the hygienic, hi-tech nursery. Even the atmosphere was enriched, with the carbon dioxide all plants need to breathe, for the sake of increased yields. Rotting straw bales provide this themselves, but for the rockwool here they had a new gas injection plant, whose bright metallic parts glistened in the sun, like a life-support adjunct to some lunar settlement module. To enter the glasshouse I had to walk over special pads which sterilized my shoes, and the environment inside, smooth and plastic, hot and stuffy, with a strangely antiseptic smell, reminded me momentarily of an enormous modern hospital, with thousands of green-robed patients standing in line with drips attached, waiting to be operated on.

These cucumbers are a variety called Corona, an F_1 hybrid type developed in Holland. To produce an F_1 or 'first filial' hybrid, first achieved with maize in the United States during the depression, the breeder perpetuates each of the two parent varieties separately. When two true-breeding plants are obtained, after repeated self-pollination, they are then crossed to give an F_1 which often shows 'hybrid vigour' and great uniformity. This basic cross must be repeated annually, however, for the F_1 hybrids will not themselves breed true. Corona, like all indoor cucumbers, is also parthenocarpic, which means that it produces fruit without fertilization. A fertilized cucumber becomes bulbous at the end, develops a bitter taste, and can leave the consumer burping over her afternoon tea. To avoid the necessity of removing male flowers, all these plants are female.

A foreman showed us the computer control room. It seemed as though the entire process was run by computer . . . the feed mixing and application, the temperature and humidity, the ventilation . . . and I was dazzled by the array of hardware. On one system the grower had 150 set points to alter. Temperatures would then be lowered and raised during the night, taking into account all the exterior weather factors. Values throughout the nursery were recorded every seventy seconds and stored on hard disc for three years, so the grower could look back and see what exactly was happening at any point. And the computer could do self-analysis, so that if, say, a certain valve had not opened properly, heating to 53° instead of 57° in a certain area for a part of the night, the computer would scan the information and disclose where there was a failure. "We should get 26,000 boxes an acre this year," the foreman told me with evident pride. Then he added graciously, "But the weather still makes the biggest impact on production. You can't do without the sun."

Each nursery grades and packs its own produce, which at Carrdale involves a great deal of sophisticated machinery. The shrink-wrapping which is now almost universal lengthens the shelf-life of cucumbers by 60 per cent. From the nursery they are then sent to the Humber Growers warehouse near Chris's nursery at Poole Bank. Here there is an incessant bustle of activity involving incoming tractor-trailers, fork-trucks pirouetting between vast cold-stores and out-going articulated lorries. This vast storage and distribution hardware is kept busy during the winter months with cucumbers imported from Guernsey and later from Spain, and with the million or so boxes of lettuce which Chris and his colleagues grow under their glass between September and January.

In the boardroom I met Chris's cousins, Philip and Nicholas Bean, his Uncle Johnny, and the marketing manager, Malcolm Revell, who joined the organization five years after it was formed, in 1966: "I used to play golf with the brothers, and poker. I was even engaged to Ann Bean at one point. So I feel like one of the family." He sits at one end of the imposing table and calmly draws on his pipe, observing Johnny Bean, one of those original four, at the other end. "I used to think there were two sides to an argument, but I soon learned that there were four. I think we view Humber Growers as a necessary evil, like the EEC."

This was certainly true of Chris. "You can get more profit marketing yourself," he told me out of the others' hearing, "if you're doing a little bit here and a little bit there, but you waste your money in the long run because unless you look at it as a mass production system it isn't economical from a growing point of view. When you harvest an acre of lettuce, that's about 7,000 boxes, and if you plump them all on the market yourself, you'll flatten it. But you've got to cut them all together because you've got to clear the greenhouse to plant the next crop. You can't do a little bit here and a little bit there because at different stages of the crop you need different atmospheres and so on, so you have to keep it all relatively the same."

Malcolm Revell defended the organization and its emphasis on specialization: "We sell to Marks and Spencer, Sainsbury, Tesco, Safeway, Littlewoods, Asda . . . 60 per cent of our total sales go through the wholesale markets, but 75 per cent of the medium-sized straight cucumbers which these stores want go through them. We have to keep that market, and we have to reduce class 2 to nothing, culturally. . . . You've got to produce for a market, not grow something and then say 'Well, where shall we sell that?' This is the only organization these stores can come to, except for Dutch importers, and say, 'Our requirement for next week is 10,000 standard boxes of cucumbers.' And we can offer price stability, whereas with the Dutch auction system their prices can go up and down like a yo-yo. There's no way that during a week, and particularly towards a weekend, these stores want to be altering prices upwards. We can quite often go for four or five weeks without altering a price." The acreage of cucumbers grown in the United Kingdom is almost static at about 400. Humber Growers have gained sales, while growers in the Lea Valley — the other glasshouse district where poor light levels favoured cucumber rather than tomato production — have lost them. But with their increased yields the Beans have also secured markets which the Dutch traditionally held. "But we can never knock them out. To put it in perspective for you, we produce about 2,200,000 boxes a year and the Dutch at their peak will produce that many in ten days!"

Chris said something to the effect that for every cucumber grown in this country we stop one being imported, and there should be more subsidy help. But he was interrupted: "Subsidies encourage over-production. They encourage the inefficient to be more inefficient." It was only a year since some of the Beans had marched through the streets of Brussels waving banners, to put an end to subsidized gas for the Dutch growers. But there was agreement that money should be available for research into more energy-efficient structures. "New glass", or more likely new structures using plastic glazings such as ICI's Melinex, may cost £140,000 per acre. But they can reduce heating requirements by two-thirds. Chris, inevitably, has even more radical ideas, and he wants to expand his computer experience from management to design.

I sensed that the other Beans sitting around the table, some sipping a drink before lunch, some scanning the day's financial papers, were not entirely dismissive of his "playing around" with computers. After all, the Bean tractor was still in production, even if it had long ceased to earn them any money. This was designed by Bill Bean, oldest of the four brothers, who had recently died. His son Philip explained: "It was for hoeing, basically, during the last war when labour was short, and it was revolutionary in that it took all the implements from behind the tractor and driver and put them in front. When it went out of production in the late Fifties, owners kept patching them up, especially the Thames Valley growers, and eventually demand grew for a replacement. A company started to make them again about four years ago and when I saw one at a show, I couldn't see a nut or a bolt that was different, after a gap of twenty-five years. They still call it the Bean tractor."

This brought the conversation in the boardroom of what has been described as one of the most progressive growing and marketing organizations in the country to an unexpected stasis. I made a note of a statement Johnny Bean had made a little earlier, which had been passed over quickly without comment from the others: "A good grower will get good yields on any medium." It occurred to me that at no point had it seemed relevant to me to ask what kind of soil they were on here.

"My father was born in that house over there, and my grandfather too," said Johnny Bean, pointing to one wall of the room. "And we end up over there." He pointed to the opposite wall, beyond which lay the small cemetery containing the graves of a dozen or so members of the family. "So we haven't gone very far, have we? Two hundred yards!" Then his round, suntanned face broke into a jolly chuckle, "What about the future? I wish you could tell me."

6

Salad Days — Organic Tomatoes on Pilling Moss

"GOD'S GRACE AND Pilling Moss are endless," according to the medieval monks of Cockersand Abbey who first raised intensive crops on the strip of flat land between the Pennines and the sea in northern Lancashire. And so it seemed to me as I hiked and pedalled across the desolate flats in search of the monastic ruins as a youngster. But it is the open prospect, blessed with the clear light of the west coast, which has led to the success of a modern glasshouse industry here. According to the nurseryman's rule of thumb, a 1 per cent increase in light gives a 1 per cent increase in crop. The growers on Pilling Moss and at neighbouring Marton Moss, near Blackpool, and Hesketh Bank, near Southport, together reap a formidable crop of tomatoes and have long surpassed the traditional inland areas such as the Lea Valley.

Douglas Blair has almost 2 acres of glass at Low Carr Nursery near Pilling, on a strip of sandy silt which runs between the sand and the peat. He is in many respects typical of the Lancashire nurseryman, with a local background and training at the County Argicultural College, growing orthodox glasshouse crops on an average scale for the traditional markets. But in 1975, in part of his nursery, he began an experiment which most of his neighbours would interpret as a step backwards into the ignorant past, while new colleagues whom he met as a result of his experiments hailed his achievements as a bold step forward offering the only viable, long-term future for food production in Britain. In the winter of 1978 Douglas came to my father's warehouse to buy Canary tomatoes, to sustain his direct markets until his own new crop was ready. He told me incidentally that, with 20,000 cartons a year, he was now the largest-scale organic tomato grower in Britain.

When I went back to Low Carr Nursery, Douglas was in the glasshouse, taking out side shoots from his tomato plants. He had won a Churchill Travelling Fellowship to study organic food production under glass in Europe, and had published a report on it which the Soil Association describes as "essential for large and small scale growers".

And he had been an active participant in several organizations established to promote organic growing and marketing, most recently reporting on cucumber production to the annual conference of the Organic Growers Association and British Organic Farmers. But he is undoubtedly a grower with green fingers, literally green and strongly tomato-scented fingers. He was clonking down the rows with metal platforms fastened to his shoes to give him a higher reach, taking out side shoots, and re-tying the sloping strings around which the plants were trained at a lower angle to allow more room for growth. He was working in a team with three others and could not afford the time to pause to talk to me, so I shuffled along in front of him, ever wary of his foot-high iron shoes, between the rows of 10-foot-long tomato vines.

The small team was working hard with an atmosphere of co-operative purpose to their labour, unlike some typical pantomimes I had seen elsewhere: "Mind out, and let me show you," the boss would say, and then plant or harvest or trim and pack his crop in a rush of enthusiasm, with great speed and precision. For all of two minutes. Then he would wander away to survey the next crew, with the expectation that his brilliant example will be followed all day, all week.

In the Blair nursery even the tomatoes seemed to grow not to order but more in a sense of benevolent co-operation. Perhaps my perception was coloured by those extra percentage points of sunlight; or because the nursery looked so smart and productive on a human rather than a technological scale; or because there is an aura of civility to the organic approach, however much it is also rooted, in this instance, in straightforward economic calculations and crude common sense. In any case, there was an atmosphere in the nursery which seemed to illustrate a quite different paradigm from that which I had seen operate in other nurseries, and which offered a quiet corroboration of Douglas's words.

"We'd been growing tomatoes for more than fifteen years, my wife and I, and just gradually came to question the things we were doing. We were putting on a basic standard feed every watering, for instance. If it was dull we increased the potash. If it was dry we increased the nitrogen. But the weather would change before you turned round. And all that water-soluble fertilizer . . . the plants were taking it up non-stop, by osmosis, whether they needed it or not. That softened the cell walls and made the plants more susceptible to disease. Then when we got a disease problem . . . well, there were sprays for everything. Each new brand that came out made grander claims than the one before, but we found that the more we used them, the more we had to."

A member of his staff who had worked at the nursery for several years interrupted across the rows of vines: "I used to spend hours just mixing up the feed . . . all the chemicals. . . ."

"And we found that however carefully we followed instructions in applying these sprays, some of our pickers would get skin irritations. We started to think that there might be a real danger to our customers, from residues left on the fruit."

Another call came through the greenery: "You see them working in other places with masks on. Growing food. How can that be 'ealthy?"

"We were down looking at a glasshouse in Evesham, for instance, where they use NFT, and the chap's mate was telling us how they were putting nitric acid through. . . ." Douglas called across to his own mate to ask what the nitric acid was for. "Anyway, it's in the feed, this nitric acid . . . to keep the pH right or something . . . and the machine had gone wrong and it put a tremendous overdose of nitric acid through and it killed most of the crop. That is, the crop appeared to be dead. And of course before anyone noticed anything wrong, all that nitric acid was going into fruit that was sold and eaten. It's a bit frightening. He said they lost five or six trusses but then the plants recovered and they got a better crop later on." The man in the next row finished the story: "He said there were nowt left but bits o' sticks with no leaves on 'em, but then they all came round and they got a better crop than they'd ever had! So maybe next year he'd do it again, and start off by going round and killing them!"

"These are the kinds of things that can go on when you're mucking around with chemicals." At this point Douglas was interrupted by his wife Penny, who is a working partner in the business. Their two teenage daughters needed ferrying to various extra-curricular activities after school, and it was almost time for a tea break. When these domestic arrangements were settled he gave me a brief account of their organic method:

"We were playing with it for two years, and then in 1977 we went fully organic. There was a lot of trusting to common sense, and respecting the soil, and seeing if it worked. According to the Ministry analysis we shouldn't be able to grow a crop on this soil at all feeding just plain water, and yet it is the soil and the compost we mix in which is the basis of good husbandry. It's the compost which supplies the humus which feeds the soil bacteria. The bacteria in turn provide the plants with nitrogen, potash and phosphate in soluble form which they can use. When the plants are taking up less food the bacterial activity automatically decreases to arrive at a balance, which means that in biological husbandry we don't have the problem of daily critical quantities the way we did with applied liquid fertilizers. The nutrients go in the soil and then we feed the plants pure water.

"Starting in February we buy 100 or 150 tons of farmyard manure, which is cheap enough in itself but is getting a bit expensive to cart here. We add very little to that: calcified seaweed, working out at about 3 oz. per square yard, which is full of trace minerals and stimulates the bacteria to keep the pile sweet. And in the past we've used a little blood, fish and bonemeal. Our main problem is getting hold of enough potash from a source that the organic movement finds acceptable. They advise using comfrey for crops like tomatoes which need high potash. But we're not talking about a six by eight foot greenhouse. We've got an area of comfrey growing outside and we cut it and compost it and we get enough for one bay, one barrowful. We'd need as much acreage of comfrey as we have tomatoes! You see, comfrey is only 2 or 3 per cent potash. There's a new rock potash brought down from Scotland, which is ground up rock with about 12 per cent potash but it takes quite a while to become active in the soil and we're not sure how good it is at the job. We're very short of research facilities. So for the time being we're using sulphate of potash which is 48 per cent potash. It's mined in Germany and Israel, it's not something which had been manufactured. We're putting a bit on at the start and then small quantities throughout the season so that it's not all available to the plants at once. The Soil Association, who set the standards and whose 'organic' symbol we sell under, are allowing us to use it at the moment, but it's under review all the time.

"Anyway, we bring in the compost during the following December, and work it into the top few inches of soil. We've got a very fine soil which recovers well after it's wet, but grinds down easily and dusts up a bit when it's dry, which is not so good for aeration. The paths used to become as hard as concrete and we'd have to come in with a cultivator and crack the paths up specially. Now we don't have to, so it's improving generally. We're building up particle size each year.

"It used to be our tradition to sow the seed on New Year's Day. We moved forward to December 6 for the last two years to try and catch the early market when prices are high but we had very big fuel bills and it hasn't worked too well. It may be that the way we grow our tomatoes just isn't suitable for that approach. We use Dutch seed and we've grown Wilset this year, but it needs too much phosphate early on and too much messing about with. Whereas Ostona, which we normally grow, seems to do well under our kind of conditions. We normally set the plants out in the nursery in February and start picking in April, going on until the end of October with the same plants. We train them up string, and when they reach the top we re-tie at an angle, like we're doing now, and then eventually nip them off. We take out the side shoots once a week until we start picking, but then it's a matter of finding the time. At the height of the season we pick the crop three times a week. It would be an easy job if we didn't have to pick. . . ."

The glasshouses have automatic irrigation, switched on by a device which measures the rate of evaporation. And they are heated by coal. The Blairs have a keen interest in energy conservation and were one of the first nurseries in the country to install the new generation of coal-fired boilers. This has made tremendous savings, but the manufacturers had to iron out various teething problems. When it comes to other innovations, such as the use of thermal screens which are pulled over the glass to prevent heat loss at night, Douglas would rather wait and let someone else sit out the teething. In any case, these new developments invariably need new structures, and the capital investment needed to erect new glass is huge.

I asked him about pests and diseases and he seemed a little embarrassed by his glib reply, hunching his shoulders: "Basically we don't get any. The plants are strong and healthy, with strong cell walls. Sometimes we get red spider in, and we buy a biological predator. It costs about £7 per thousand. The other headaches, the things like brown root rot, eelworm and Fusarium wilt that build up in the soil, we avoid by sterilization, just like every other grower round here. You force steam through pipes that you run under the top few inches of soil, after the crop is cleared in November, and you kill off everything. It's a shame we have to do it, because it wipes out all the bacteria we've nurtured that season, and we've got to start again with the new compost we bring in. But with all your eggs in one basket it would be a terrible risk, not sterilizing. Crop rotation is the best way out of it, so that specific pests can't accumulate, and now we do grow nearly a quarter of an acre each of cucumbers and capsicums. But they're on a different scale and not so heated, not so early, so I think we're a long way off getting a proper rotation established."

The Blairs sell most of their tomatoes and all other salad crops through their own delivery rounds to shops in the Preston area and the Lake District, including a North Lancashire supermarket chain. Any surplus is now being sold to organic wholesalers, although in the past they have been sent to the traditional wholesale markets of Leeds, Sheffield, Bolton and Preston. Those wholesalers may not have appreciated the Soil Association symbol and what it stands for, but they did appreciate that the Blairs' fruit was fuller flavoured than average, and that it would stand in good condition longer. The Blairs' returns often showed a small premium above the market average as a result, and there was a demand for their produce which they could not meet. Now their tomatoes are accompanied by point-of-sale materials such as leaflets, posters, labels and price tags, all boasting the advantages of the organic husbandry which produced them. "It's hard work," Douglas says of this aspect of his business, "but very satisfying."

Douglas Blair believes that the vast majority of tomatoes on the market give consumers very little satisfaction in terms of quality. A Soil Association writer has described them as having "the texture of moist cotton wool, the thick skin of a pop star and the blandness of blancmange". They are engineered to the growers' and salesmen's criteria, and they neither look, feel, smell or taste like real tomatoes. Nor is Douglas necessarily impressed by high yields. "I grow just short of 100 tons an acre and there are growers getting 150 tons. But it doesn't mean a thing. I'm trying to persuade people that yield alone is no longer an important criterion. What's yield anyway? I mean all right, these growers on rockwool or NFT may have twice as many tomatoes as I have, but that's not what counts. Quite apart from the quality, the real quality of the fruit, it's whether or not the job pays at the end of the day. And if it costs them twice as much to grow twice as many, what's the point and where's the advantage?

"Take the big companies down south," he says, "who employ the most sophisticated technical equipment and controlled growing methods. They go massive and get big gluts and what do they do? They start shipping them up here! All they do is create problems for everybody else."

Unwanted gluts of a more natural kind may have dominated the earliest of tomato crops, for some botanists believe that tomatoes first made their impact as field weeds of maize and beans in Central and South America, much as rye and oats originated as weeds of wheat and barley in the old world. A member of the Solanaceae family which includes the potato and capsicum, the tomato grows truly wild in Ecuador and the Galapagos Islands. It was probably taken into cultivation about three thousand years ago by the skilful farmers of one of the member nations of the Inca Empire. The tomato naturalizes very easily, its seeds passing undamaged through the digestive tracts of most animals and germinating readily where dropped (vast quantities of "wild" tomatoes formerly grew in sewage farms, and many gardeners find them sprouting from kitchen waste in their compost heaps). From the Incas' fields the tomato passed via Mayan

Guatemala into Aztec Mexico. It was from Mexico, after its conquest by Cortes in the early sixteenth century, that the Spaniards first brought seeds to Europe, and it is from the Aztec word *tumatl* that our common name for the fruit is derived.

From Spain a Moor took tomato seeds to Morocco, and in the mid-sixteenth century Italian sailors calling at Tangier took the fruit to Italy. Over the next hundred years tomatoes were gradually accepted in southern Europe as vegetables for cooking, but botanists and other writers were suspicious of them, and in northern Europe they were classified as strictly ornamental plants well into the nineteenth century. An early Italian writer called the tomato *Mala insana*, the unwholesome fruit. A Frenchman called it *Lycopersicon*, wolf-peach, an Englishman subsequently adding *esculentum*, edible, to give us our modern botanical name. Other experts decided that the tomato was an aphrodisiac: "It gives little nourishment to the body, and that little bad and corrupt." And so it became the *pomme d'amour*, the love apple. Elderly people in rural areas were distrustful of the fruit even into this century.

Towards the end of the last century, however, the commercial nurserymen of Britain became attracted to a fruit which was so easy to grow and which cropped so heavily. The early glasshouses belonged to the kitchen gardens of wealthy estates, producing mainly peaches and apricots, melons, figs and grapes. In 1840 the first commercial greenhouse was built at Worthing, but the industry which rapidly expanded from there still concentrated on luxury foods, developing an expertise in grapes, in particular, so that Britain was not only self-sufficient but actually exported them to France and the United States. In 1849 a Mr Slaughter, of Worthing, records a buyer from Bond Street collecting 40 lb. of tomatoes every fortnight, but he also remembers ripe fruit dropping from the plants, because there was nobody who would buy it. Growers persevered, however, and with a season now extended from weeks to months, and with prices steadily dropping, the tomato became increasingly popular eaten raw in salads, so that by 1904 a writer for *Country Life* ventured to suggest that "when the knowledge of its culture is acquired by our artisans and labourers with gardens in country districts, it will become an article of daily food among them".

A vast glasshouse industry developed to serve London in the market garden districts around Epping, Cheshunt, Waltham Cross in the Lea Valley, and tomatoes were the primary crop. By 1935 there were 600 growers with 6,000 employees working 1,000 acres of glass in the Lea Valley and dominating the market at Covent Garden. Prior to the development of steam sterilization in the 1920s, the problem of disease accumulation in the soil was tackled in a desperate fashion. In contrast to the Parisian market gardeners, who when pushed further afield by the expanding city loaded their carefully nurtured top-soil on to carts and took it with them, the Lea Valley growers used the same soil for their tomatoes until it became 'sick' and then took it away in great cartloads, and imported fresh soil.

Much of the Lea Valley land has in turn been sold for building development, and the pollution is now so severe as to limit glasshouse production to cucumbers. The major tomato growers are now concentrated near our southern and western coasts. Research and development continues apace. H. J. Heinz, for example, with their vast appetite for tomatoes, have recently joined with the oil company, Atlantic Richfield, to engineer genetically a new "vegetable monster" which will be a "heavier, heartier and faster-maturing 'super' tomato". They have even gone back to South America to collect nine species of wild tomatoes to breed from. Research is conducted into every possible facet of crop environment manipulation, and especially into refining techniques for growing on rockwool and NFT, but not on soil. The quality of the soil and its possibilities features less and less in the literature.

"The argument", says Douglas Blair, "is that there aren't enough of us in the organic movement to justify spending money at research stations. Well that's their way of looking at it. I have a lot of colleagues in the Organic Growers Association who are very interested in what we're doing here. But they tend to be young men with great ideals and not much money. They are what you might call highly undercapitalized."

Alan Schofield is just one such undercapitalized young man with ideals, and he is now working with the Blairs on the 3 acres of land outside the glass at Low Carr Nursery.

"I started on a dairy farm in Devon, going to college on day release, and after six years' involvement in standard agriculture and horticulture I got fed up with using all the sprays and everything, until it reached the point where I just wasn't prepared to do it. Then I got in touch with Douglas through the Soil Association, and he said he had a plan for a bit of spare land he had, and this was it."

Alan is growing a wide range of outdoor vegetables to organic standards, to supplement the direct sales of glasshouse crops to local retailers. He is very enthusiastic about what he is doing and talks with an excited confidence. "There really is a demand for what we grow. We've got the van round to the local supermarket chain and good greengrocers. And people come to us. We get all sorts of enquiries. People ring up because they have allergies, or they're dying of cancer or something and they've been told to eat organically grown foods. They're really desperate, wanting to know where they can buy them. It makes so much more sense to grow what people really want.

"I'm still a bit waterlogged, because this plot has been down to pasture for so long that the soil is really compacted. And I can't break it up by growing potatoes because of the risk of them spreading blight to the tomatoes. But you just see what I can do by working it after a year or two, with good compost!" It was already impressive, in fact, with prolific growths of healthy-looking winter Brassica, neat rows of onions and leeks putting on good weight, and long beds full of courgettes and flat lettuce, all putting the lie to the old image of idealistic vegetables full of insect holes. "The lettuce earns the most money, but it's very labour intensive. It's tempting to grow exotics, like kohlrabi, Florence fennel or endive, when the price looks high, but I want to be able to provide ordinary people with traditional vegetables such as cabbages, which they can rely on for being healthy. . . . Once I'm on top of it, one man should be able to keep this going, but I need better management on the field scale. I had a half-acre vegetable garden when I was twenty, but I've got to manage this carefully, so I can turn the land over more, and crop more. I want to work in green manures like rye grass and winter tares to improve fertility, in between the cash crops. It takes good management. Douglas has guaranteed me money to live, and then when the business is on its feet we will split the profits. I want to prove that I can earn a decent living. I reckon I can turn over £8,000 a year here, when I'm established.

"At the moment I'm buying and adapting pre-war machinery that comes up at farm sales, like mechanical weeders and ridgers. This whole area was contracted by the Ministry of Defence to grow vegetables during the war but now it's reverted to dairying. And anyway, it's all chemical equipment these days, not the mechanical type we need. There isn't one seed crop in the United Kingdom that isn't herbicided."

We were called in from the field when the tea was mashed, and Alan took me aside into a little room off the packing shed. Here he regaled me with the kind of proselytizing information for which the *Soil Association Quarterly Journal* has become a noted source, interspersed with direct assessments of his own. Last year, for example, over 4 million lb. of pesticides were sprayed on to our planet, and the rate is growing at 20 per cent each year. Residues have even been found in the tissues of antarctic penguins. "Nature is being raped all along the line." Ninety-seven to 99 per cent of all our vegetables and cereals are sprayed with one or more pesticides. Some vegetables receive as many as 15 different chemicals, and in one reported case an astonishing 46 applications in a season, yet residues are hardly monitored and there are no legal limits. "It's absolutely bonkers." Last year over a billion gallons of pesticide-carrying spray were applied in Britain. Over 20 per cent of that is released in droplets of such a small size that they will not settle, but stay suspended in the air, and drift. Many pesticides are directly related to the nerve gases banned even from warfare by the Geneva Convention as a hazard to humanity. "It's plain immoral." And even if a pesticide is banned from use in the developed nations, it is often still manufactured for export elsewhere. Apart from thousands of deaths each year from accidents in their use, an American study found that over 15 per cent of dried beans from Mexico, and nearly half of imported coffee contained pesticides which were banned from use in the USA. "It's unbelievable what people will do." And that's only pesticides. . . . There are the long-term effects of artificially produced fertilizers, the run-off of nitrogen into our rivers and water supply, the structural breakdown and erosion of soils. "We're tottering on the brink. These soils just won't support crops in fifty years. It's as simple as that."

Alan drained the last of his tea with a grunt of satisfaction and slapped the empty mug down on the table. "Luckily there's some people like Douglas around, showing that you can do something about it. You see the hospitals and doctors' surgeries packed with people of all ages, but it's not really up to the doctors. Health isn't the doctors' responsibility. It's the farmers'." He hitched up his trousers and went back out to work.

Douglas had rooted out some of the more positive, optimistic literature, and showed me the *Organic Food Guide*, which lists the main retail outlets for his produce, as well as hundreds of other organic outlets throughout Britain. According to the Soil Association, organic agriculture will cover 20 per cent of total production by the end of this century. The University of Wales is in the process of establishing the first degree course in alternative/organic agriculture. And the Minister of State for Agriculture has ordered an investigation into the research needs of organic growers.

"The local ADAS chap is interested in what I'm doing," Douglas added from his own perspective. "But only out of novelty. At least now if I have a problem he knows not to start recommending this spray or that spray. The Organic Growers Association lobbies the Ministry people all the time, and they respect us a lot more now. They appreciate that we're not a load of idiots. They know what we're trying to do. It's going to come, slowly . . . even if only from the point of view that farmers won't be able to afford all the fertilizers and the rest of it for ever more."

7

The Better Berry — Strawberries at Wisbech

WHEN MIDDLE-INCOME consumers from suburban estates make room in their freezers amidst the microwave-ready packaged meals, and drive in their second cars to the 'Pick-Your-Own' farms which surround almost all conurbations, they are establishing a sympathetic link with their pre-hominid ancestors. Venturing away from home to gather the nuts and berries of wild plants was an annual routine of genus 'homo' long before the 'sapiens' variety developed and predominated. But the family expedition to gather elderberries or sloes is now a part of our folklore. And over the last twenty years so much of our hedgerow has disappeared, while that which remains is severely cut back by machine so early in the year, that even the nutting and blackberrying outings down country lanes with the kids are almost a thing of the past. Yet the gathering instinct remains, and is exploited, seemingly to everyone's satisfaction, by the Pick-Your-Own operations.

The wicker basket may have been replaced by the car boot, but the children are still there, and the backache, and the sun and the earth, and the hours of topping, tailing and hulling once the harvest has been carried home. The gatherers may suffer a moment's doubt, as they finally lower their bounty into the freezer, as to whether they have actually saved money. But an unquantifiable factor in this equation must surely be the satisfaction — a kind of sympathetic reward — from practising again one of the most ancient acts of economy and survival.

The focal point of most Pick-Your-Own holdings is the strawberry bed. There are about 1,000 growers with 4,300 acres of strawberries, providing about a quarter of our total commercial harvest, waiting each summer for the hordes of city folk and the car boot. Strawberries need good humus and a slightly acid soil, but these conditions can be created on an intensive smallholding almost anywhere in the country.

Large-scale commercial production has developed where it can capitalize on favourable criteria other than a proximity to freezer-owning populations. In Hampshire, along the Tamar Valley between Cornwall and Devon, and in Somerset's Cheddar Valley, sheltered south-facing slopes are ideally suited to raising an early crop. The building of the railways meant that the berries could be moved quickly and smoothly to distant markets, and several strawberry growers in these areas can trace their origins to men who came to lay track, bought smallholdings, and put down roots.

The railways were also crucial to strawberry production in the Wisbech area which includes parts of the Isle of Ely, Cambridgeshire, west Norfolk and south Lincolnshire. Here the numerous horticultural smallholders found the strawberry ideally suited for their available family labour, and for giving a quick pay-off when planted under new fruit orchards. But above all, the flat silts of the fens provided highly fertile soils with the important ability to retain moisture. "Drop a strawberry runner on the land around Wisbech", according to a local saying, "and it will grow". While, historically, Kent has provided most of the strawberries for the London and southern markets, the Wisbech area has supplied the great cities of the midlands and the north, as well as the processing industry.

Harold Selby has grown strawberries near Wisbech most of his life, and in retirement has written a booklet describing this business through the first seventy years up to the 1950s. "Land preparation in the district varied widely, but the basis of success in early years was the great fertility of newly ploughed grassland after many years of grazing by sheep and cattle. The great sheep fairs give an indication of the humus building up in the soil in the nineteenth century, twenty thousand or more sheep going through the King's Lynn auctions at a main fair. By the time I took an interest in strawberries the magnificent fertility of the fens had begun to fade, and the best growers then practised heavy feeding with farmyard manure, and other organics such as wool shoddy or bonemeal while 'London muck' from the stables of London, carried to us by rail, was widely used and cost very little above the rail charges. Most growers kept a number of pigs in yards, specifically for providing manure for soft fruits rather than for profit."

Wisbech is called the Capital of the Fens, and was a major port and trading centre before King's Lynn. But in the Middle Ages the rivers Nen and Ouse silted up at Wisbech eleven miles from the sea, and eventually the Ouse made a new course to its present estuary at Lynn. In 1480 Bishop Morton of Ely initiated the first major drainage project since Roman times, with an artificial dam to concentrate the course of the new river Nene through Wisbech. Then in 1630 the Earl of Bedford commissioned the Dutch engineer Cornelius Vermuyden to drain the fenland systematically. The resulting network of dykes, canals and embankments, with sluices to keep back the high tides, drained the area so well that the peat soil of the southern fenland shrank. The drains became higher than the surrounding land, so that expensive pumping installations and constant vigilance are required to prevent flooding. But an area which had once been a vast expanse of unhealthy and uncultivated bog, except for a few small islands, was converted into some 300,000 acres of the most fertile land in England. And Wisbech was reborn as a prosperous market town; as a trading centre shipping corn and oil-seed rape along the coast and to the continent, and as a port importing coal from the north-east and timber from the Baltic. But not yet as a cannery for strawberries.

A certain Doctor William Butler declared one idyllic summer's day, while Wisbech still contemplated its rescue from the bog and he beheld his bowl of strawberries: "Doubtless God could have made a better berry, but doubtless God never did." A great deal of the mystery and the awesome respect for nature may have been swept from farming since those days. But there is still a fickleness to every season which tends to inhibit even the most arrogant manipulators of the earth and her bounty from claiming that they can better the creation of God. And yet the strawberries for which Wisbech is now famous are very different from those which formed the "rural man's banquet" of Doctor Butler. They are in fact a quite recent innovation, and owe their creation at least in some part to the contrivances of man.

The strawberry which grows wild over most of Europe, and which was gathered by our ancient ancestors, is *Fragaria vesca*, derived from the Latin *fragga*, meaning fragrant. It is smaller than a blackberry with a rich flavour but a rather dry texture and, as well as a food, it has long played an important role in herbal medicine, and cosmetics. This plant was taken into gardens in the Middle Ages and its fruit improved somewhat by selecting the best progeny and enriching the soil. The Alpine strawberries, such as Baron Solemacher, are specific varieties of *Fragaria vesca* thus improved, and the Hautbois are similarly derived from another, less common, wild European strawberry, *Fragaria elatior*.

An early traveller in America discovered "the strawberries grew so thick that the horses' fetlocks seemed covered with blood". (The Strawberry Fields of Lennon and McCartney may have evoked equally strong images for another generation, but in fact relate to an orphanage in Liverpool so named.) In the seventeenth century, this scarlet woodland strawberry of eastern North America, *Fragaria virginiana*, was brought to Europe. It produced fruit of a slightly larger size than the native berries but of poorer flavour. Cross-breeding was not possible, as the European species were diploids, having chromosomes in pairs, and the American was octoploid, with sets of eight. Hybridization was genetically impossible, although some market gardeners no doubt suffered their ignorance of this fact with years of frustrated hopes.

Then in 1780 a French naval officer blessed with the auspicious name of Frézier, brought to Europe another octoploid strawberry *Fragaria chiloensis*, which he found on the island of Chiloë off the coast of Chile. In fact this plant was native to the entire west coast of America as far north as Alaska and may have been cultivated by the Incas. The great mountain ranges of the Andes and the Rockies had prevented cross-breeding between the two American species, but this was quickly achieved in Europe. The French botanist, Duchesne, raised Ananas and the Englishman, Michael Keens, produced Keens' Seedling. These were the first of the modern large and luscious strawberries, which far surpassed all previous varieties in both size and flavour. One of Keens' varieties was taken to USA where it served as parent to Hovey, which largely created the strawberry industry there, and others were introduced to France with the result that for many years large-fruited varieties were known as *Fraises anglaises.*

By 1892, when Thomas Laxton bred the famous Royal Sovereign which is still grown today, there was a proliferation of varieties, and by 1909 a record area of 30,000 acres in Great Britain were devoted to the crop.

During the First World War only strawberries picked and marketed on Saturday could be sold to the public, the rest being made into jam, largely for the troops. By 1919 the acreage was down to 15,200 and though this improved subsequently, a second recession occurred in the late Twenties caused largely by problems with disease. Harold Selby tells this story in his Wisbech history:

"The twenties saw a swarm of new varieties, largely from the Continent. Strains of Paxton had depreciated, but any runners were in demand, and disease problems really got going; so growers were ready for any improvements. I dropped right into these troubled waters. Madame Kooi was one of the first varieties from Holland.

Coarse in the leaf, rather more hairy than usual, an ugly shape and a useless flavour. But how it cropped! Ten tons per acre in places, if only for a year or two." Madame Lefevbre had rather sparse leaf on a long petiole so that on a hot season the fruit tended to cook on the plant, but a prolonged flowering allowed it to crop 2 tons even in a frost year. Around 1930 Oberschliesen was introduced from Germany, giving a 4-ton crop against the 2 tons of Sovereign. "Ober had a pale unattractive colour and for a genetic reason the somewhat rounded berry failed to pollinate completely so that it often grew a greenish 'nose'."

Selby lists many such varieties, each with its own advantages and flaws; each enjoying a passing popularity. With Huxley, the transience extended even to the name. It was raised by Albert E. Etter of Ettersberg, California, and brought to Britain in 1912 under the name of Ettersberg No. 121. It was subsequently called Huxley Giant, Evesham Giant, Evesham no name, Chipfiller and Brenda Gautrey, among others. Selby recalls how a neighbour invited him to look at a fine new variety in 1929: "The beautiful single plant looked much like Huxley but Mr Gautrey was certain it was a new seedling, and its crop rate was very good indeed. He said he would name it Brenda after his daughter. Thus it was grown by a number of Wisbech growers very successfully. But finally the Royal Horticultural Society was asked to make a decision, and pronounced that there was no botanical difference between Brenda and Huxley."

The great variability in strawberry yields, and the constant search for new varieties, was due to the strawberry's susceptibility to a range of devastating pests and diseases. Selby lists the strawberry tortrix caterpillar (the only control, lead arsenate), the elephant bug (which could cut the flower crop by 75 per cent) and the seed-eating weevil (certain fields just couldn't grow strawberries because of it) as three of the worst pests. But reduced vigour because of virus infection, devastation by fungi causing red-core disease, and loss to the mould Botrytis in wet seasons became so severe that some growers forecast the end of the bulk-producing business. This situation was rescued in the early 1950s by work at the East Malling research station, where potted plants were grown in a temperature of 98°F for one or two weeks. This heat killed many plants, but a few healthy ones survived. The Nuclear Stock Association was formed to grow and distribute the virus-free clones resulting from this heat treatment and there was a vast increase in yield. Then in the early Sixties chemical sprays against botrytis rot were introduced. The generation of varieties which accompanied these developments, many raised by D. Boyes at

Cambridge University, proved more reliable and Cambridge Favourite provided over half of our total crop for many years. Cambridge Vigour, Cambridge Rival and Cambridge Late Pine are also still widely grown.

Strawberry plants are very sensitive to length of day. At a latitude with a certain length of day a given variety may produce many runners but no flowers and fruit. At a different latitude it will produce the flowers but no runners, and at a point in between the two it produces both flowers and runners. Much of the recent breeding work has aimed to develop the Remontant or perpetual varieties which grow few or no runners but ripen fruit from July to October. These newcomers are often launched with extravagant claims in the gardening press, but the Wisbech grower is not enthusiastic. He needs a flush of fruit for volume picking, and when the short hectic season is over he is generally glad to see the back of it. "June and July are the months for strawberries," one of them told me definitively.

Harold Selby captures some of the hectic quality of the season when he describes the transport of the fruit, once picked: "There were the problems of the fitness and readiness of horses; a cast shoe on the road was a minor catastrophe; the rush to take perhaps three loads in a day to keep to the railway time-tables which had to be closely adhered to; the occasional canter to the station in an emergency instead of the usual trot. Station yards closed their gates late in the afternoon and there was little sympathy from local stationmasters if your last load was shut out, though there were ways of increasing this sympathy! Every inch of space in yards after a hot day's picking was taken up by horse vehicles of all kinds, with a long queue outside the gates, and a wait of two hours was not uncommon. Fast fruit trains comprising sometimes 30 or 40 specially ventilated well-sprung box wagons left the district daily in the height of the season for the marshalling yards at March or Peterborough to be attached to express passenger trains. Markets as far as Glasgow and Cardiff would be reached by six a.m. following loading at three p.m. the previous day." Anyone who has endured the notorious trials of travelling across East Anglia by public transport in recent years may enjoy contemplating the fact that in Harold Selby's youth a regular service for Wisbech strawberries was provided to the large cities by fast train from stations at Wisbech, Upwell, Coldham, Elm, Emneth, Smeeth Road, Gedney, Wisbech St Mary, Guyhirn, Holbeach, Postland, Clenchwarton, Long Sutton, Magdalen Road, Manea, March, Middle Drove, Cowbit, Murrow, Ferry, Outwell, South Lynn, Spalding, Sutton Bridge, Tydd, Terrington, Thorney, Middle Drove and Walpole.

In my own youth, lorries were sent by the wholesale merchants to collect strawberries on the farms and holdings. It was a long trip home to Lancashire leaving Wisbech at teatime. This was the pre-tacograph era when drivers carrying perishable produce had special leeway with their hours but grew tired at the wheel however fresh the load behind them. I especially remember Brownhills, on the A5, receiving an unscheduled delivery when one of my father's lorry drivers hit a lamp-post, escaping unhurt himself, but tipping 500 trays of strawberries on to the pavement.

Collecting the load could be tiring, too. Selby talks of special trains from Liverpool Street on Saturdays in June, laden with families with crowds of children, cooking pots and bedding, going to Wisbech to pick for strawberry growers. But Wisbech always had a large contingent of smallholders using their own family labour — over 4,000 growers were listed during the Second World War, most of them with less than an acre. My father's drivers would call at a dozen or more holdings to make up their load, sometimes driving through a council estate, where a man's ¼-acre "back lot" could provide an important boost to his income from regular employment elsewhere. Some fruit would command better prices than others; all had to be tallied and accounted for, and by Saturday a host of separate cheques were written, to be mailed to farm cottage and semi-detached in virtually every housing district of Wisbech.

Roy Handley's 25 acres lie three miles out of Wisbech along the road to Downham Market, at Elm. Top fruit is his main business. He has no cold store and all his commercial apples are sold from the trees through Norman Hargreaves & Sons Ltd. The smalls go for cider. He grows a little sugar beet, which is easy and profitable. And he has 2 acres of strawberries, a ½ acre of which is always newly planted with non-bearing maidens. Last year he marketed 8 tons of berries through Hargreaves, and 3 tons went to the processors. "There's plenty of us left in the 10- or 12-acre range," he says, "but the ½-acre men are going fast."

"I buy my plants from a special plant grower at King's Lynn. They've been Cambridge Favourite for twenty-five years now. Before that there was Brenda, and lots of others. The plants go in thirty inches apart each way, over a ½ acre. I plant in October or November, but I don't get any fruit off them in the first year. We put herbicides on at planting, though we can't use the ones we like because they're too strong. There's always hand weeding. It's a lovely soil here, quite silty. You could grow anything on it. But there's lots of weeds. You can't put any of the good herbicides on after Christmas, or the kind ones on after flowering, or they'll stunt the growth. I don't put straw down the first year. I go down between the rows with a cultivator once a week and pull all the runners up into the rows, and then when they're thick enough we cut the rest off."

For the following three years, Roy Handley puts straw down in late May. This acts as a mulch against weeds but above all it keeps the berries clean and dry, up off the soil. Our name for the fruit is derived from the Anglo-Saxon 'streawberige', a word with obscure origins owing nothing whatsoever to Roy Handley's, or anyone else's, relatively recent annual labour of spreading straw. Some hold that the name derives from 'strea' berries because the runners cause the plants to 'stray'; others that it comes from an old practice of threading berries on straws of grass for easy carrying by children, or from the achenes or seeds on the outside of the berries, which look like the chaff of straw.

"I spray four times against Botrytis, during flowering, and for mite and red spider. Then I have about 10 pickers in for the 1½ acres. Some are local women, regulars, and some are school kids who've finished their O or A levels. I can tell which are up to it straight away, and I can have them back again for the apples. You get smaller fruit in the last year. The same tonnage but smaller fruit, so that's plugged and sent to the processors. Then you can chew up the runners and rotovate the straw in. I cut all the tops off the plants with a grass mower and that makes spraying easier next spring."

This description sounds very mechanical, but there is a fanciful note, conveyed more in his expression and gestures, which says: 'This is the strawberry lark and I'll try and explain it as best I can.' The strawberry lark has much to do with long, hot days in the open sun, with sweat and beer, and the liquid presence of ripe fruit; with deep colour, rich fragrance and the fullness of time; with all the heady dreams and hopes of summer. It is also, of course, about money.

"Strawberries is the biggest headache of all. Two or three years back we were wiped out. It's a terrible headache. But it's important because it's early money. It's some cash coming in early, after a long winter and spring."

I asked Roy about new technologies for strawberry growers.

"Not that strawberry picker! It's a waste of time. I've seen it demonstrated, but it's only for processors. It's like a grasscutter, it cuts everything off and blows the débris away and then another machine with contra-rotating rollers pulls the strawberries off. All the little green ones go in, too. But it doesn't matter, I suppose, because then they add colouring. In the winter I go to ADAS meetings to hear about work on new varieties, or to a show given by one of the companies, to launch a new chemical. You can learn from other growers as well as the experts, and it's a nice social occasion."

I asked him about Kentish Garden, the country's largest softfruit co-operative with grower members far afield from Kent. Roy had never heard of it. "We're very independent round here. I don't believe in co-ops."

I told him that I had read a report about Kentish Garden using sophisticated pre-cooling and temperature-controlled storage facilities, and his distrust deepened, visibly. And as well as extending the shelf life, they were also extending the season: with production under polythene in the early months, and new late varieties for the tail end, they could extend the season from May to September. So there was now an outdoor, volume crop for almost four months of the year. But that broke the spell, and he burst into a cheerful smile.

He could perhaps have made fun of bloated, tasteless, out-of-season strawberries, shiny red with their preservative spray, which are air-freighted from the far side of the world through the rest of the year, at a price. Instead he said with a grin, "Ah, well. That's as may be. But for me one month is enough."

8

Veg from the Vine — Peas in South Yorkshire

MIDWAY BETWEEN THE heavy industry and manufacturing towns of Doncaster and Scunthorpe, near the village of Epworth, there is an unassuming sign at the side of the A161 bearing the title, J. A. Gagg & Sons, Lawn's Farm. Several driveways leave the main road in the vicinity of this sign. One of them leads to a large yard surrounded by enormous sheds for storing roots and grain, fertilizer and other chemicals, trucks, tractors, harvesters, and all the 'plant' associated with the running of a hi-tech farm of 4,700 acres. Down this driveway for six weeks in the months of July and August, three or four lorries operate a non-stop shuttle, twenty-four hours a day, seven days a week. In those six weeks, come rain or shine, day or night, 1,000 acres of green peas are harvested and shelled, and the lorries rush them thirty miles to a freezer factory in the one hour of grace before the load turns sour. It is a harvest which seems to have more in common with the rolling imperatives of the nearby sheet metal works than the timely appropriation of nature's bounty.

Another lane near the farm sign leads in a sweeping circle to the modern bungalow, in the suburban American 'ranch-house' style, of Michael Gagg and his young family. Michael is an energetic Yorkshireman in his thirties with an evident good sense of humour, an enthusiasm for his work, and a cheerful willingness to talk about the growing and harvesting of peas. He insists that he is a 'farmer' rather than a 'businessman', 'salesman', 'administrator' or any of the other words which describe essential roles on a farm of this size. And he has the robust air, and the wind-blown, sunkissed face of a man who works out of doors. But he brings none of the dust and sweat inside. He pours me a glass of lager and relaxes in his low-swept modern furniture. In the yard he raises pheasants for shooting, but his other winter sports are enjoyed at the family villa on the Spanish Mediterranean.

"Yes, it's a big farm. Bigger than most around here. We put up the initiative and did things that other people didn't want to do. We took a lot of risks, taking a couple of farms and then borrowing money for the others. Most of it has come in the last fifteen years. Father only started with 90 acres.

"There's quite a lot of growth in dried peas in this country, because they're very high in protein, and some of them round here grow peas for drying. Mushy peas are canned dried peas, and there's a surprising amount of them eaten. And they also use them in animal feeds, because all the protein for feeds, like soya and tapioca and so on, has to be imported. The dried peas are bought from farmers by merchants called 'pea-pickers' because it's their job to take out bad peas. They're called 'hand-picked' if they're for human consumption, even if they use machines, and the rejects are called 'pea pickings' for animal feed. . . . Then up towards Doncaster and Leeds they grow pulling peas, for selling fresh, because they can get people out of the cities to do the work. We used to do that once, but we can't get the people out here nowadays. So we grow what I still call garden peas, but all ours are frozen. They're what are called vining peas.

"We drill from the first week in March to the middle of May, using different varieties to get a good sequence at harvesting. We used to grow Kelvedon Wonder years ago but they're out-of-date now as far as we're concerned. We grow mainly Avola, Scout, Pugit and Small Sieve Freezer, which is a very dark green colour, part of the family which started off with DSP, or Dark Skinned Perfection. These are grown specifically for freezing because we want a small pea. When you get some big Kelvedon Wonder on a plate, they don't look so attractive. We want small peas with lots in the pod and double or triple pods on a branch.

"We aim for 14 plants per square foot over the whole field, so we're putting on 16 or 18 stone of seed per acre. The seed costs £50 to £60 per acre, so it's quite an expensive crop to sow. With cereal we put on about 14 stone per acre and it's about a quarter of the price. The peas are drilled with five or seven inches between the rows, but they fill the field so that within a few weeks you can't see any lines.

"We usually give the field 50 units per acre of phosphates and potash each, which is about 2 cwt, before we drill. But no nitrogen at all. We can kill the weeds with one spray, using a pre-emergent before the peas come up. Then we could possibly have to fly-spray them later on, but that depends on the weather and what sort of action there is of greenfly. But we usually spray only once, if we have to do it at all. We put traps up to handle pea moths, so we don't get those little grubs inside the pods. There's a special thing like a little triangle about a foot long and we put a sticky card in it coated with one of the hormones, either male or female, and it attracts the other lot. We put two in each field of about 45 acres and keep a close eye on them, changing the cards when they need it.

"Then before you know it, it's vining time, and that's when you know the meaning of work. We hire a few extra staff to help with other jobs, because it's peak time, but our own men cope with the vining. They work twelve-hour shifts and they love it. Because of the overtime."

He sighed and gave a wry smile. "I always say the viners are designed to break down.

"The pre-viner device started off as a static machine in the yard in the 1950s, when either pulled peas or peas with all the straw were taken to it. Then in the mid-Sixties we started taking them into the field and towing them around with a tractor. We'd get stuck then if it was too muddy. Then in about 1978 the pod-picker came into existence. With the old trailed pea-viner you used to have to put all the pea — the haulm and all of it — in, and it used to be a very slow job taking it all through. But with the pod-picker it's a lot quicker and one machine has actually replaced two of the old trailed viners and the cutter. Now the viner strips the leaves and the pods off the plants with rotating fingers at the front that go through like a comb. Then they go into a big drum at the back which fetches all the peas out of the pod, by rubbing the pods together, which splits them open. It's a big circular thing, like a net, and the peas drop through like a sieve at the bottom, and all the rest — what we call the vine — comes out the back. The peas can easily suffer damage, so you've got to set them up very well. It's done by putting a certain amount through the drum, and if they're damaged too much we'll let the back of the viner down to let them out quicker. Or alternatively we can slow them down if we're not getting them all out of the pod. It's all done by hydraulics. The viners we use are made by FMC, which is actually an American corporation, based at Fakenham in Norfolk. They cost £120,000 each.

"Still, the viners are very temperamental. They're designed to tilt, to go up and down hills in the Yorkshire Wolds, for instance. It's the same principle as they've got now on the Advanced Passenger Train. Before they ever ran that train we said we could tell them all the things that would go wrong with it, and we were right! . . . The trouble with our viners is that when we start them going, we don't stop. We're working twenty-four hours a day, all day and all night, seven days and seven nights a week, and when they break down we take our own workshop with us, out into the field. . . . Even if it pours down we keep going, although we're getting to the point where if we have very bad weather we've got to stop because of problems it would cause for the land.

"Peas are a good crop for the land, generally. The beauty of the pea crop is that it's got this nitrogen-fixing property on its roots. It takes nitrogen from the air and fixes it in the soil, and this natural nitrogen that doesn't leach out is just right for the wheat that follows. So peas are what we call a good break crop. After the peas we grow wheat, which we don't come to until the autumn. Sometimes after early peas harvested at the very beginning of July we've put in a few turnips afterwards, for sheep feed over the winter, but that's on particularly light land where you could keep sheep. And as well as cereals we grow a lot of carrots. And we've 350 beef cattle that live inside all the time, fed off carrot waste and other arable by-products. We can only grow peas in a given field once in every five or six years. More often than that and you don't get such good yields. We don't know whether it's because of a pea eelworm, or whether it's got to do with the bacteria which are left in the ground from the nitrogen fixing, or what it is. But the land gets sick of peas.

"As far as other nitrogen fixers go, we're too far north here for ground beans, and runners are for market gardeners, they're not suitable as a field crop. We used to grow a lot of broad beans, in fact we started off with them instead of peas, but the broad bean market has become much smaller, and you've got to grow them very near the factory because they've got to be washed very quickly once they've been vined, or else they get stained. We're about thirty miles from the factory, and that's far enough, even for peas. They come out of the drum slightly bruised and wet and will get hot, just like your compost heap, and go sour. We've got to get them there in one hour.

"We freeze our own peas under contract with the cold stores of Christian Salvesen. They started in Scotland, making money out of whaling. That old station in South Georgia that triggered the Falklands War was one of theirs. All the big processors started on the east coast to deal with fish, at Hull, Kelso, Felixstowe, Yarmouth and so on, and that's how Grimsby became the largest vegetable processor in Europe. Now Christian Salvesen have got cold stores all over the country. Women pick over the peas before freezing, and then we bring them out and pack them in the supermarkets' own labels and so on as we want. But I'm a farmer. I try to keep out of the other side of it. We have an agent who handles sales. Some parts of the industry are . . . well, it's a matter of knowing people all the time isn't it? Like everything . . . Birdseye and Smedley have their own factories so we're competing with them to a certain extent, competing at the retail level. We use the same kind of discipline as these major companies impose on their contract growers. We're members of the Pea Growers' Research Association, which has done a lot of work on rotation and things like that, so we have the same sort of information as Birdseye have got. And the market for frozen peas is static. There's a sale for about 100,000 acres a year. If we get about 5 per cent of acreage above that the whole market crashes."

Historically, the market for garden peas has been far from static. A vegetable deemed appropriate for peasants in Germany constituted a 'royal dish' specially presented to Henry II of France by his Italian wife. Garden peas were sold in the streets of London in the reign of Henry VI but in Elizabethan times were specially imported from Holland, being "fit daintie for ladies, they come so far and cost so dear". And in the seventeenth century, fresh peas swung out of favour, from fashionable delicacy to vulgar commonplace, in short order. Madame de Sévigné wrote of "Impatience to eat them, the pleasure of having eaten them, and the longing to eat them again," while a contemporary of hers remarked: "It is a frightful thing to see persons so sensual as to purchase and eat green peas."

Peas were one of the earliest vegetables to be brought into cultivation, some remains at an archaeological site in Burma having been dated as preceding 9000 BC. Peas have been found at a Stone Age site in Hungary, and a Bronze Age Settlement in Switzerland, and they were mentioned in Homer's *Iliad*. They belong, like beans and lentils, to the legume family, characterized by their distinctive five-petalled flowers and their seeds being carried in pods. After the Graminae, or cereal family, the legumes, because of their high protein content and ability to fix nitrogen in the soil, have been perhaps the most important group of plants grown for food, both in the old and new world. According to the Book of Genesis, Esau sold his birthright for a "pottage of lentils"; lentil remains were found in a number of Egyptian twelfth-dynasty tombs (before 2000 BC); and there is an original Sanskrit name for lentils, which indicates that they were grown in India before 1500 BC. There are similar records of broad and field beans in antiquity, although the debate continues as to whether these legumes originated in southern Europe or central Asia, and of chickpeas, soya beans and other beans or 'grams'. Similarly, the kidney beans and scarlet runners not introduced to Europe until after the voyage of Christopher Columbus had been cultivated in central and southern America for over 3,000 years.

There are eight amino acids, or proteins, which adults cannot metabolize in their digestive systems, and ten for children, which are therefore 'essential' if protein in the diet is to be 'usable' or 'complete'. Animal proteins contain all the essential amino acids but, with the exception of soya beans, plant proteins are deficient in some. Cereal grains and legumes, when eaten together, make up for each other's deficiencies to provide complete protein, and food historians have speculated that the vigour of, say, the Incan, Indian and Egyptian civilizations was based in part on the balance provided respectively by maize and kidney beans, dal and rice or chappati, and lentils or chickpeas with bread or tahini in their national diet. It is also the case today in a world increasingly aware of its limited resources, that it takes on average 10 lb. of plant protein to produce 1 lb. of meat protein.

Another complementarity is at play in the field, where the cereals' need for nitrogen to be readily available in the soil is met by a crop of legumes planted there in the previous year. Nitrogen is very inert and much energy is required to separate the two atoms in its molecule. A bacterium called rhizobium enters the roots of legumes as a parasite, stimulating them to make nodules in which the bacteria then proceed, using the trace element molybdenum as a catalyst, to separate nitrogen molecules from the air and combine them with hydrogen to form ammonia. This is then 'fixed' in the soil awaiting its take-up by the cereal crop which follows. Bacteria in these nodules will continue to fix nitrogen even after the plants are harvested and cut, as long as the root stalk is left in the ground. When the old two-field system of rotation, planting half the land with cereal and leaving the other half fallow, was replaced by the three-field rotation of cereal, legumes and fallow, in medieval Europe, farmers had both better yields and better protein. One writer has speculated on how important this innovation was to the emergence of Charlemagne's empire.

Certainly, it was in the Middle Ages that dried peas became a common constituent of the British diet, with pease pudding eaten hot, cold, and nine days old. And it was after the Norman invasion that we adopted the word 'pulse' for legumes, from the Latin word 'puls' meaning pottage. Our word 'pease', with the plural 'pison', derived from the Latin *pisum*. But in later centuries, 'pease' was mistaken for a plural, giving us 'pea' singular and eventually a new plural in 'peas'.

Pease pudding was made by tying dried peas, perhaps with some chopped herbs or a little spice, inside a linen pudding-cloth. This was then suspended from a hook in an iron cauldron of boiling water, which might also contain earthenware jars full of beef, mutton or fowl, with eggs and onions, as well as bacon or other pieces of salt meat wrapped in linen to hold in the fat and juice. After several hours in the boiling water the pudding would be lifted out of its bag like a hard green cannon-ball, crumbling into lumps when attacked with a knife and making a fine meal with the salt meats.

When peas are allowed to dry, much of their sugar content is converted to starch, which accounts for their great thickening properties, and they also develop more protein. If dried peas are subsequently encouraged to sprout, by keeping them warm and moist, the starch becomes the energy source for the manufacture of the vitamins and other essential structures which the 'seeds' need for growth. After several hours of germination, peas develop a high vitamin C content which was entirely lacking in the dried state, and after two days sprouting they become an excellent source of B vitamins. They also taste delicious, with a crisp and crunchy texture.

Whether or not medieval cooks took advantage of the mini-harvests represented by sprouted dried peas, it seems likely that they appreci-ated the brief period in which peas could be eaten fresh. This season even became associated with courting, for however rare fresh peas may have become by Elizabethan times, earlier generations believed in "winter for shoeing, peascod for wooing". If a girl placed a peascod, or fresh pod, containing nine peas above her door, the first man who entered was to be her future husband.

It was not until this century, and the bulk canning of peas, that the short season for the vegetable was extended and they became available as 'garden peas' all year round. Bryan Donkin was the first to set up a tin can factory, at Dartford, Kent, in 1812, but for many years canned foods were considerably more expensive than fresh and were bought mainly by explorers, or by those who found, say, truffled hare or woodcock indispensable to their survival in far-flung outposts of the Empire. The cans were heated prior to sealing, in the belief that driving out the air preserved the contents. Usually this killed any harmful bacteria and the contents were in fact preserved. From about 1870 canneries were aware of the work of Louis Pasteur, and in the USA machine-cut cans replaced hand-made ones. On the domestic level the Americans soon lived by the aphorism: "Eat what you can, and can what you can't." Internationally, their new mass production led to huge exports. In 1925 all the tinned peas consumed in Britain were imported. Ten years later, however, S. W. Smedley had a factory canning 250,000 tins of English peas per day, or 9 million tins per season.

The Americans also pioneered the preservation of food by quick freezing. The Chinese stored winter ice for summer use, in special houses kept cool by evaporation, as early as the eighth century BC. And the Mughal Emperors successfully brought snow from the Hindu Kush for their summer sorbets in Delhi. But both depended on a natural cold source. It was not until the 1850s that the first ice-factories applied the technique of evaporation to highly volatile materials such as liquefied ammonia, and not until 1929 that Clarence Birdseye set up an industrial process for food preservation by freezing at Gloucester, Massachusetts. The industry did not become truly established in Britain until after the Second World War, but we now grow 100,000 acres of peas each year for freezing, against 6,000 acres for eating fresh.

Many market gardeners believe that there is still a profitable market for fresh peas as a seasonal luxury on the strength of their superior flavour. And by the same criterion they are increasingly growing more exotic crops. Varieties such as Waverex are particularly suitable for picking when still very young and small. Marketed as 'petits pois', these are especially sweet and tender. Other varieties, such as Oregon Sugar Pod, are picked when the peas are barely formed and marketed as 'mangetout', 'sugar peas', or in the United States 'snow peas'. Sliced for stir-frying or eaten raw in salads, these pods are crisp and juicy, and don't leave a wad of fibre to be chewed. Those labelled 'snap peas', such as the variety Sugar Snap, have even thicker, fleshier pods.

Certainly few would claim that frozen, and even less so canned, peas preserve the flavour of fresh. Many also argue that the breeding of peas for vining — by shortening the flowering period, for example, so that more peas are at the optimum stage of development at the same time, or by producing semi-leafless varieties — has been at the expense of taste. Norman Asquith, who markets fresh peas in West Yorkshire, is not one of these, however:

"I think our vining industry is the best in the world. And peas grown for freezing *are* very nice for eating fresh. They're beautiful peas. They've got to be, or folk wouldn't go on eating so many." He has grown fresh peas for many years, but now markets only those of other growers. "He does a very nice little package of fresh peas," a retailer told me. "And they sell well."

"I used to grow 200 acres. There was a big trade in peas once upon a time, in 40-lb. hessian bags. Whole families would sit around shelling them into a bucket for dinner, and they'd have to whistle to stop from eating them raw.

"People round here still grow Feltham First, which is an early, Laxton Progress Number 9, Early Onward, and Onward. Most people have stuck to them. These new varieties eat very nice, but seed is terribly dear. And it's like anything else — if you like it you don't change. We used to grow Alderman on sticks, but you couldn't afford it now. So growers plant rows every four and a half inches, instead of the twenty-eight inches in the old days, and by the time the peas are ready they can't see their land, it's just a mass of green and that sort of holds them up.

"Like I said, I don't grow peas any more. But I market peas, and there's still a good trade for them. I believe there always will be. There's a woman I know who buys a 20lb. bag of peas every week of the season and she says that's all that will keep her husband in. He takes a basin into the front room every night and sits there in front of the TV and pods them and eats them."

Some growers have dropped out of the business because of the difficulties of finding pickers. As one of them explained to me, "The casual labour comes out of the cities, so they're not local people who we know. We'll go in a field in a morning and collect names to withhold tax, and find that we have twelve Mickey Mouses, twenty-four Margaret Thatchers and so on. It's a problem." But with fresh peas fetching better prices on the market, the growers who remain are more optimistic. "If the public wants fresh peas, it has to pay for them. You can use machines to cut the vines, but not to pick the peas off the stalk."

9

The Iceberg Cometh — Lettuce on Rixton Moss

"OUR LETTUCE IS a life-enhancing product, and smoking is not life-enhancing," says a successful Canadian grower as he routinely stops visitors with cigarettes from entering his greenhouses. But he says it with a pleasant smile, according to reports in the trade press. His 350,000 head of lettuce per year react to human moods, he believes, and grow less well if people around them are bad-tempered.

The growing and marketing of lettuce in Britain is experiencing several new developments, and "leading the revolution", according to the Fresh Fruit and Vegetable Information Bureau, "is the iceberg lettuce". This outdoor lettuce, grown in tremendous quantities in California's Salinas Valley, is also being marketed as a life-enhancing product. "With the growth of fast foods and salad bars, and a greater awareness of healthy eating, more people are now eating more salads than ever before — around 25 lb. each every year, up $1\frac{1}{2}$ lb. from ten years ago." But growers are presumably encouraged by the market's long-term potential. Our average consumption of $2\frac{1}{2}$ lb. of lettuce per person per year has a long way to go to reach the Californian equivalent, which is 28 lb.

The interest in iceberg lettuce may have been kindled by the large numbers of Britons who visited the United States in the 1970s and were enamoured with this crisp, sweet salad which came with every hamburger, pizza and taco. My own initiation came in the company of the Californian artist Juana Alicia, and was rather different from the norm. Juana had worked in the lettuce fields of Salinas and would not describe either the process or the product as "life-enhancing". Nor did the lettuce grow in an environment of pleasant human moods and good temper. Juana suffered painful skin allergies as a result of the intensive application of chemicals to the crops. Others suffered severe respiratory problems, and pregnant workers had a high rate of miscarriage. The migrant Mexican-American labourers were ex-ploited in a manner reminiscent of the Twenties and Steinbeck's

Grapes of Wrath. And the efforts of Cesar Chavez and the United Farm Workers Union (UFW) to improve their lot met with increasing violence from the rival Teamsters organization. Juana's striking mural 'Las Lechugueras' (The lettuce workers) for the City of San Francisco shows these workers — strong and purposeful and yet so vulnerable, with the foetus of a pregnant woman no more protected than her hands and lungs — toiling down the long monolithic rows of lettuce behind the mechanized harvesting rig, with the plane flying over and spraying all indiscriminately. I joined the UFW-sponsored boycott of iceberg lettuce, resisting the lure of its bland sweet taste (incidentally, compared with $2\frac{1}{2}$ lb. of lettuce, we already consume, on average, our own weight of sugar every year). Instead I bought locally-grown green-leaved lettuces which had not only a flavour of their own, but also the higher mineral and vitamin cotent and the aura of genuine freshness which for me deserves the attribute of 'life-enhancing'.

One of the pioneer British iceberg growers sought to answer my doubts about the chemical-intensive cultivation of American-type icebergs: "We have found that the lettuce keeps better if applied fertilizer is kept to a minimum. We only use a tenth of the quantities common in California." But this says nothing about pesticides and herbicides. British growers made many trips to the United States in the late 1970s to study the Californian system of production.

So instead I went to the Lancashire Moss growers, to write of an agricultural endeavour which echoes an earlier American experience — that of the first pioneers — and a lettuce crop which from my point of view is more salubrious and instructive.

As glaciers flowed down from the Pennines during the great ice age they plucked out, ground up, and carried along with them, huge quantities of boulders, sand and clay. As the glaciers merged at lower levels, they spread out into vast ice-sheets, and as these sheets melted around 10,000 years ago, they plastered the lower Mersey valley with an uneven layer of boulder clay. On the poorly drained level ground between Manchester and Liverpool, a chain of hollows in the clay became shallow lakes. Sedges and reeds gradually filled these shallow lakes, so that by five thousand years ago they had become fens. Decaying vegetation became waterlogged and sank to the bottom, becoming compressed and partly carbonized over thousands of years to form peat. Alders grew in the swamp, but as the minerals needed for most plant growth diminished, sphagnum mosses gradually predominated, and the moss peat slowly built up to form the dome shape of raised bogs. These peat bogs, or 'Mosses' . . . Chat Moss,

Barton Moss, Trafford Moss, and a host of others with varying local names through the ages . . . were wild, treacherous places, avoided by travellers. In his tour of England in 1724 Daniel Defoe wrote: "the surface looks black and dirty and is indeed frightful to think of, for it will bear neither horse nor man. What nature meant by such a useless production it is hard to imagine, but the land is entirely waste."

A small area of Risley Moss has been retained as an educational nature reserve, whose achievements deny the description of the land as being "entirely waste". As for the rest of the Mosses, in most cases they have become highly productive agricultural areas.

Rixton Moss, a five-mile stretch of country between Warrington and Irlam, separated from Risley Moss by the line of the Manchester railway, was brought into production as recently as 1879. At a time of agricultural depression when many farmers' sons were leaving Britain for better prospects in the Canadian prairies or the Australian outback, a small group of pioneers left home to try and wrest a living out of this small area of wild, virgin land — "a crust of peat on which heather and stunted birch struggled for existence above many feet of slimy ooze" — which lay all of fifteen miles from the Free Trade Hall in the centre of Manchester.

I went to visit David Mawdsley, the great-grandson of one of those pioneers, and his uncle Jim Austin, who still farm on their families' original holdings on Rixton Moss. With the help of some of their old press cuttings and oral histories taken by Rangers at Risley Moss, I pieced together the background of what Jim Austin calls "the best lettuce in the world":

"Someone brought to the village of Tarleton, near Ormskirk, a tale about the bogland at Rixton. It was said that the owner, Lord Winmarleigh, was prepared to let plots of it at a nominal rent to young farmers willing to tackle the job of cultivating it. Most of the farmers shook their heads, but others were inclined to let their sons take a chance. Papers were signed, plans were drawn and boundaries pegged. Then one day in 1879 a group of young men set out from their homes with a few hand tools and other possessions piled on borrowed farm carts. There were young Blundell, Dandy, Johnson, Marshall, Mawdsley and three or four others, the oldest only twenty-five, the youngest twenty; sturdy lads who had learned farming the hard way. Each was allotted a plot of 20–25 acres on which to test his skill and hardihood. And tested he was.

"The first few months were spent on combined operations — making roads and digging ditches . . . working from dawn till dusk. . . .

"The main thing was to get it dry. The drains was the main job.

"It was possible to thrust an iron rod eighteen feet into the bog by hand, and consequently when digging the drains it was only possible to take out one graft — that is, the depth of a spade — at a time, and lay it on the side of the drain for a week or so before taking out the second graft, until the drain was about three feet deep. This long drying-out process was necessary to allow the sides of the drain to consolidate, and thus avoid caving in.

"The drains would be about eight yards apart, feeding into giant drains every hundred yards that would become the boundaries of the fields and carry the water down into the Mersey or its tributaries. Unlike so much fenland, we are above sea-level, which makes all the difference.

"The fight against the bog involved long, back-breaking hours but it seemed to be a healthy life. Most families were big, and people lived to a good age.

"There's a story about one of the Marshalls who lived in a house made of turf sods which was so damp he had to sleep with his matches under his pillow to get a light in the morning. Some farmers lived in old railway carriages, or wooden railway sheds . . . or any kind of wooden shack that would sit light on the peat.

"Old tram horses which would walk in a straight line and did not easily panic could be bought for as little as a shilling per leg. All the animals had to be tethered, for if they were allowed to wander they disappeared, not to be seen again until the bog yielded up their bones. Early on, the horses had to have pattens — twelve-inch square wooden platforms fastened to their hooves, like snowshoes — even in their stable, if they were in the middle of the moss.

"Then they had to sweeten the soil, because a peat bog is acid, and that meant spreading marl (clay) in place of lime. They might put 80 tons on an acre, all with a hand-shovel and fork. The marl would give weight to it, too, and there were marl-holes all round the moss when we were kids. They'd fill with water and we used to go playing round them.

"For nitrogen they'd put manure on, and nightsoil. Lord Winmarleigh built a little siding to save on haulage and they'd bring in thousands of tons of nightsoil from Manchester. It was very cheap because there were no sewage works — people had closets at the end of the yard and they were emptied at night by the council workmen, and the Corporation couldn't get rid of the stuff. We also had a muck wharf down there on the river. They'd bring all these cattle from America and so on and they'd put all the muck they'd made in the ships on to barges and bring it up here and you'd have a stint getting it out, and throwing it on the bank and then into carts, all by fork, and that was a job. . . ."

The same railway siding soon began to carry produce away from the moss. "They put a dozen gruelling years into their land before they made it pay. Wiseacres told them to go back to Tarleton before the bog broke their hearts. But they all won through except one, whose health failed." At first potatoes and peas were the main crops, but at the turn of the century lettuce and celery became popular and very profitable, and some bunching salads like radish and spring onion which was the women's work and brought in a little money early in the season. Each grower would have two Irishmen from mid-July for the main-crop celery that was grown trench style; they'd soil it up and so on and they'd be paid so much for twenty yards. "It was graft, real graft.

"We'd supply Manchester and Liverpool and some Yorkshire markets. Then according to my mother it was 1921 or '22 when they had a real dry summer and everything was burned up in Bedfordshire and the Thames Valley. There was no irrigation there in those days and so there was no lettuce in London. But a group of buyers put their heads together and came up here, because the peat will hold its moisture in a dry year, and they bought our lettuce, and we've been sending to Covent Garden ever since.

"There was some flat lettuce grown but it was mainly Cos and we used to send so much of that to London that we call it London lettuce up here. And in London I believe they call it Manchester lettuce! This area still grows a third of all the Cos lettuce in the country. We can grow anything on this soil, but our lettuce is the best in the world because it grows so fast that it doesn't have time to go leathery. It's as crisp and tender as you could ever have. It melts in your mouth.

"I can remember these Webbs coming in just before the last war . . . Suttons Al was the first variety we grew . . . and you could hardly give it away, they just didn't know what it was. Then it gradually started selling in Yorkshire. Our fathers always said they were canny in Yorkshire. And now there's all this talk about iceberg, which is only Webbs really, wrapped up with money. And people think these iceberg lettuce co-ops are some kind of fairy godmother. But I don't fancy it."

Green Alley Farm house, with foundations sunk down to the underlying clay, was one of the first brick houses, built to replace the wooden shack of pioneer William Mawdsley. I walked from there up

Holly Bush Lane, past cow-hole field whose name still recalled some distant tragedy with the farm cow, and others full of lettuce, celery, carrots, cabbage and leeks. The soil was like a fine powder, difficult to walk on, and soon covered my shoes and pants with black dust. My hands were suddenly dirty, and then I started scratching my head, so that by the time I reached the old railway siding I felt as though I had been swept along an enormous Victorian chimney.

David Mawdsley says the moss grows anything, perfectly. "But we don't grow cauli's, because at a dry time like this you get a lot of dust floating around and once it settles on the curd there's no way it comes off. I've seen us trying to wash it out, and you just can't. We don't get the strong blows here like they do in the fens, but you can get a lot of dust floating around from time to time. The peat is contracting all the time, as it weathers away and shrinks as we drain it more. Some old timers can remember just seeing the roof of the signal box up there, and now you can see the bottom of it. It must have come down 40 feet in their life. You can touch clay with the plough in places already, but there's twelve feet of peat in other places. It varies a lot. We're not really worried yet, but in the long term it'll be a problem." And then he laughs and adds sacrilegiously, "When we're down to bare clay we'll have to build houses on it."

Green Alley Farm had 57 acres when I visited, and about 35 of them are down to lettuce, although some of these are cropped twice in the same season. An acre of moss land grows 2,400 dozen lettuce at its best. Two-thirds of David's lettuce fields have been planted to Cos varieties, although there is a swing towards growing more of the round Webbs types.

Cos lettuce was first brought to England from the Greek island of Kos by crusaders. Wild, bitter-tasting lettuces from which our crops were originally derived grow over most of Europe, western Asia and northern Africa: the ancient Greeks certainly cultivated lettuce as a salad, but it may not be much older than that in cultivation. Our word derives via *Lactuce* from the Latin *Lactuca*, *lac* meaning *milk*, for the milky juice in the stem of the lettuce plant. If not cut in time the lettuce head will become 'blown', opening to grow a tall stem bearing yellow flowers and then seeds with a pappus of soft white hairs, which act as parachutes and aid dispersal in a wind.

David Mawdsley buys lettuce seedlings ready-germinated in blocks for the very earliest crops. Then he sows his own seed in progression, starting in a greenhouse in early February, and then a cold-frame, until by the end of March he can drill his seed directly in the field. He still uses 10 or 15 tons of farmyard manure per acre for celery but he is turning increasingly to chemical fertilizer for lettuce. "It's getting very expensive in labour, transporting and spreading manure, and then clods of it tend to come up to the surface and foul up the inter-row cultivator. And it brings in a lot of weed seeds." Weeds are a big problem on the moss, especially fat hen and pickle weed. They are attacked with chemical herbicides, the inter-row cultivator, and the hand hoe. "People are getting more conscious of nitrates being left in the plant from fertilizers we use — Heinz analyse the celery they buy off Glazebury Moss, for instance, and won't buy it if the nitrate level is too high. Within five years we'll have statutory analysis of random samples. With the pesticides there's a time limit. So we might spray once a fortnight for greenfly, that's two or three times per crop, but we'll stop three weeks before we cut it."

David grows Lobjoits Cos, followed by Dark Green and then Valmain for the very late Cos-type, all of which form a good heart without tying the leaves round with raffia, as was done in the past. For the round, Webbs-type he grows Penlakes, Saladin and then Great Lakes, and is always trying new ones. Saladin is widely grown for iceberg lettuce, but it likes perfect growing conditions and if it doesn't get them there are problems.

The first lettuce is ready near the beginning of June, and they are cutting from the last sowings into October. All this work is done by hand, by David and his brother Gordon and their four fulltime employees. They have a selection of razor-sharp kitchen knives of all shapes and sizes, bought at the local flea market, and everyone has their own preference. Three casuals are hired in the summer months, chiefly for packing. Then the lettuce goes to the same markets as it did in his father's day and his grandfather's — in many cases to the same family wholesale businesses. But where they used to pack 160 dozen lettuce into each railway truck at 'Winmarleigh station' they now pack 1,400 dozen into a single lorry for the quick journey down the M6 and M1.

David typically spends the hour between six and seven a.m., when his workers arrive, on the telephone to his markets. If his wholesaler tells him that lettuce is making 120 pence per dozen, he can reckon to deduct 10 per cent salesman's commission and 5 pence handling charge; 39 pence for packaging and 29 pence transport, which leaves him with 35 pence before he begins to consider his labour and production costs.

The overheads for iceberg lettuce are on a scale of their own: a shrink and over-wrapping line, for example, might cost £25,000; a vacuum cooler, £60,000; or a laser-controlled land-leveller, allowing more uniform planting and more uniform growth by regulating the water table beneath the crop, £95,000. Expenses of this magnitude could only be met on the Moss by growers forming a co-operative.

The families on Rixton Moss shared a pioneering experience almost within living memory. Many of them have intermarried. The Risley Moss Rangers describe it as a very close-knit community. But this word 'co-operative' is something else. Jim Austin has lived and worked on the Moss for sixty years, and it makes the hairs on the back of his neck bristle. "You can't sell your own lettuce yourself, for one thing. You've got personal contacts with your outlets that you've had all your life — and you don't easily give it up. Then another thing I object to is the amount of grant you get to do it, to set it up, and the intervention grant that co-ops get, such as with cauliflowers when the price is down, that individual growers don't . . . because I'm one of these people, and David's probably very similar, who believe that a business should stand on its own feet, and not be government aided."

With or without the Mawdsleys, however, a co-operative is being formed on the Moss for the growing, packaging and marketing of iceberg lettuce. The public appear to want this product, and it can command a premium price. It has been predicted that within a decade this type of lettuce will account for most of our consumption.

Iceberg lettuce is not strictly a variety, but rather the product of a method of production and packaging. Its name derives from the days earlier this century when large crisp hearts of cabbage-type lettuce would travel in railway trucks from California to the east coast cities, packed around with huge chunks of ice to keep them fresh. Now a large number of 'crispy' or 'Webbs-type' varieties are available, such as Saladin, Lake Nyah and El Toro, which, under careful cultural conditions and given reasonably good weather, will grow an especially large, dense heart. The yield per acre is lower, because of greater spacing and because up to 20 per cent of the crop may be left in the field to maintain a high-grade, uniform product. And all the looser, outer leaves are left in the field on cutting, so that the product is typically six inches in diameter and $1\frac{1}{2}$ lb. in weight, and is 100 per cent edible.

Specially built rigs housing up to twelve workers are towed by tractor alongside the cutting crew. As the rig crawls along, the women on board tidy the lettuce and then within seconds wrap it in moisture-retaining film which still allows it to breathe. At the end of the rows the ready-boxed lettuce is transferred on pallets to another vehicle and driven to the packing shed. The loaded pallets are wheeled into a vacuum cooling cylinder which removes the field heat, cooling the lettuce to 1°C very rapidly. The lettuce is then kept in cold store, transported to market in cold-chain vehicles and often to super-markets with refrigerated display stands, so that it stays cool right up until the point when it is tossed with the salad dressing. It may have cost four times as much as a traditional outdoor lettuce, but it will keep in the salad tray for up to ten days. And the tightly interlocking leaves will stay together when cut into cubes, wedges or slices. Technology has given us a clean, easy lettuce which we can keep readily at hand. As one iceberg grower is on record as saying: "If there is a demand and people are prepared to pay a realistic price we will meet it. We would be happy to cut our lettuce standing on our heads in the moonlight if people wanted to pay us to do that."

10

Pressed for Taste — Garlic on the Isle of Wight

———————————————

THERE WAS THE Beatle who dropped out of the group just before the first bestselling single, and there was the partner of Allen Lane who left publishing just before Lane launched his Penguin imprint.

This ex-publisher's son became a farmer in Kent, and then in 1958 moved to Mersley Farm on the Isle of Wight. He ran a small dairy herd and grew maize as a fodder crop. Tempted to sell the ripe ears as fresh vegetables, he soon learned the difference between field corn and sweetcorn. By 1961 he had 2 acres of sweetcorn and a contract to supply fresh corn-on-the-cob to American airbases in the south of England. The acreage was doubled every year. He developed his own KingCob label and designed a new packaging system, with some of the husk stripped off to reveal the corn, which presented the ears at their best. He made a deal with the Milk Marketing Board for his corn to be sold alongside their butter. Then he became the exclusive supplier for Tesco supermarkets. Soon he was growing 40 per cent of the total crop of the United Kingdom.

Martin Boswell has more-or-less retired from farming, and the market for fresh corn-on-the-cob which he pioneered has much expanded, with many farmers across the south of England muscling in. KingCob may be the largest volume single brand, with a reputation for high quality, but it now commands only a small percentage of total sales. Martin's son Colin has become the chief executive at Mersley Farms and he has inherited much of his father's style and approach. He talks of "becoming a big fish soon", not with arrogance but with evident pleasure. In fact he is already a big fish. The crop which has earned him this status is one of the most ancient in cultivation and yet it has been held by recent generations to be so antithetical to the British palate that it is perceived by most of us as something new. It is the most pungent member of the Alliaceae, *Allium sativum*, or garlic.

Colin Boswell seems, on a brief acquaintance, to be eminently British in a solid, modern sense. And yet he rustles a tray of dried garlic bulbs — banned, according to a recent belowstairs report, from all the Queen's kitchens — with obvious delight. "That's how it should sound. There's a good feel to it, too. The vegetable business is changing rapidly. There are a lot of new crops. We're growing early potatoes and carrots under plastic, because we can beat even Cornwall then. We grow a little asparagus. And then we've some late calabrese for October and November. But I've got a special attraction for garlic. There's a certain aura to it. No other crop has the same magic."

I am impressionable enough to believe that there is a certain aura to farmers who talk like this: like the men with big rough hands who hold a head of lettuce like a newborn baby, or a handful of potatoes like cut-glass. Perhaps it is the special combination of pride and awe; of a profound respect and simple acceptance. And yet in Colin Boswell this aura comes not with a romantic rustic who chews on straw or ploughs behind a horse, but a marketing expert in smart clothes with a radio-telephone ever at hand.

"I've still got grower's hands, but only just," he admitted. "I like to work in the fields but the chaps will tell me to sod off and do what I'm best at. I keep in close touch, though, and we've got radios in all the offices and tractors. It's like planning a campaign. We had to choose the right variety to begin with and then plan how it should be planted and cultivated, harvested, dried, stored, cleaned and packed. The workforce reaches as high as 150. It can become a strain being both manager and farmer." A foreman interrupted to relay the plight of 20 tons of garlic from France which has been delayed by a dockers' dispute at Dover. "Seventy per cent of our garlic is sold through supermarkets. Tesco was the first to take it, along with the sweetcorn. We are also exclusive suppliers to Marks and Spencer, and we sell to Sainsbury. They want quality and continuity. And they are cutting down on the number of their suppliers. So we import foreign garlic from February to July. It comes from all over the world . . . Spain, Italy, Mexico, Taiwan, Argentina, Chile . . . the French import it in 2,000 ton loads and sell off 20 tons to us. Then we re-pack it." This garlic is sold in the same KingCob boxes, but overstamped 'produce of Spain' or whatever. "You get to compare quality firsthand," Boswell says with a smile of pride tempered by the hurt to his pocket. "There's always more wastage with the imported bulbs. We called in the Gas Board once to investigate what we thought was a leak, and the villain turned out to be 15 tons of Egyptian garlic."

Most of Boswell's garlic is packed two or three bulbs in a see-through box, with twenty of these to a carton. But he hires two French ladies with nifty fingers to braid bulbs into traditional plaits, which are then sold in nets with loops at the top for hanging prettily in distinguished kitchens. He sends two trucks to the mainland each week, by British Rail ferry, and can reach every wholesale market in Britain within twenty-four hours. For Boswell believes that there will always be a place for the independent retailer. He wanted to know what kind of garlic was available in the greengrocer's shop in my own small town's high street. "This generic advertising is all very well," he laughed, "but then we'll have everyone thinking they're on to a good thing growing garlic and we'll get swamped the same way as we did with the corn." Then he ushered me into a Land Rover and bounced me down miles of rough farm track on the edge of Mersley and Arreton Downs to take a look at the garlic in the field.

The Isle of Wight measures only thirteen miles from north to south and yet encompasses in that distance seven strata of rock running east-west across the island, each giving rise to distinctive soils and land use. The parish boundaries are based on those of the Roman manors which, though narrow, ran north-south the entire length of the island, giving each holding the entire range of different soil types. A strip of chalk runs across the centre, from the Needles to Bembridge Foreland, and between here and the uplands at the very south of the island lies a band of Lower Greensand, cross cut by the East Yar river and its numerous small tributaries, with strips of modern river deposits. This is a fertile, well-drained region with clear air and a great deal of light; with mild winters and early springs. It was this area which earned for the Isle of Wight, before Kent, the title "Garden of England", though local farmers have not always sought to capitalize on that particular potential of their land: Paul Hyland (in *Wight: Biography of an Island*) describes a work dedicated to "all lovers of husbandry, particularly to those of the Isle of Wight", who were then berated for their ignorance, idleness and insularity. They were rebuked for resisting the new developments of the eighteenth-century agricultural revolution: "We must not mistake the meaning of Solomon where he says, 'There is nothing new under the sun', that being metaphysically spoke relative to the knowledge of God."

Garlic is undoubtedly new to the Arreton valley, but it thrives under the Arreton sun. "We can plant out in November, as the winters are so mild," Boswell told me as we rattled along under the Downs. "And these small fields give us some protection against the westerly winds. We can plant by machine, but we get better results by hand.

The seed is very expensive, about £1,000 per acre. We use cloves from
the previous crop as our seed, not seed from the flower, and these
cloves have to be broken from the parent bulbs by hand. It's very
laborious. By harvest time a crop of garlic can cost £1,500 to £2,000
per acre, which is a great deal more than most crops. That, and the cost
of the drying rooms, explains why every other farmer won't be
rushing in." We hit a series of pot holes and he spoke louder as he put
the Landrover in a lower gear. "We started with ¼ acre and now we
grow about 150 tons on 40 acres. In the long term we should follow a
twenty-year rotation. This is because of the danger of nematodes
building up in the soil. They can be present still after fifty years, but
the population drops if there are no hosts. We really need virgin land
for the garlic. We own 380 acres and farm about 500." The field he was
taking me to was leased from a neighbour. "The cloves are spaced five
inches apart, with thirty inches between the rows. We buy in cow
manure from neighbours, but we also put on a little nitrogen,
phosphate and potash. We weed through the winter, using some
chemicals and hand hoes, and we spray against mildew. The bulbs
grow well with the early spring. Garlic is a biennial really, but by
removing the thick stalk we get all the strength into growing a big
bulb, which develops entirely underground. We harvest in July, when
the leaves turn yellow and lean over."

We arrived at a field of some 5 acres, dotted with workers and stacks
of wooden trays. "We sometimes have a 10-acre field with 70 workers
in it. That must be a rare sight in modern agriculture." As if to press
home the point the radio-telephone beeped and Boswell put it to his
ear. A youngster wanted work. Boswell asked his age and told him to
come to the office the following morning. "We get labour from all
over the island. There are a lot of youngsters who've left school or
college and they're looking around for what they really want to do in
life and this kind of work suits them fine in the meantime."

This was true up to a point. The young men and women who were
crouching or kneeling in the soil, clipping the tops off garlic which had
previously been lifted to dry in the sun, were cheerful enough. They
wore colourful T-shirts advertising the Penn State basketball team or
the University of Missouri at Columbia, but they were all "caulk-
heads", or islanders. They were paid piece-rate by the row and all
whom I spoke to were satisfied with the money: "I wouldn't want to
do it forever, but it's okay" . . . "the smell's not bad," one of them
added, "but it's not like strawberries is it? I mean there's not much joy
in nibbling as you go along is there?" I learned later that the
unemployment rate on the island was 16 per cent, compared with 10

per cent across the Solent, and that the elderly "overners" who retire to the island are balanced by the youngsters who emigrate.

The air was redolent with the smell of garlic. An aura, indeed. The bulbs looked ripe and full as they sunbathed on the south-facing slopes of the Downs. Colin Boswell mixed amiably with his workforce and eagerly rustled the clipped bulbs in their trays as he passed. They were enormous compared with the products of my own Shropshire garden, and a lovely clean bright white, tinged with pink.

"Each bulb has to be snipped and cleaned," Boswell explained, "and after partially drying in the field it goes into the drying store where it's kept at 80°F for several weeks. Then we can put it on to the market. As long as we keep it in a controlled atmosphere we can sell this garlic in peak condition right through until February or March. We could keep it longer, but once released from store in May or June it would sprout very quickly. That's when we use imports."

Garlic is in demand all year round, and the demand is growing. Common Market statistics reveal that more garlic is now sold in the south of England than in northern France.

An enveloping breath which reeks of garlic is an obligatory ingredient of the British caricature of the Frenchman. A French salad dressing can seem like a feeble attempt to dilute garlic with a little oil and vinegar. Garlic is at the core of every sauce, every casserole. It finds its way into their bread, their cheese and even their butter. They coat their cooking utensils with garlic. Balzac, that appreciative French gastronome with few inhibitions regarding self-expression, suggested that the cook himself be rubbed with it. And yet the myriad uses of garlic were exploited long before things French.

The crop was first cultivated in the Kirghiz, the mountainous region between Afghanistan and Mongolia, where it is still found growing wild. The Chinese word for garlic is written with a single sign, which means that the plant was significant there when their written language was being developed. Garlic is mentioned in the Book of Numbers in the Bible, and according to Herodotus the ancient Egyptians made great use of it. The first recorded strike of organized labour was supposedly resolved only when the 300,000 slaves building the great pyramid of Cheops were granted an increase in their daily ration of garlic.

Many species of Allium have played an important part in adding a zestful flavour to bland staple foods, down the centuries and throughout the world. Some Egyptian cults made a god of the onion. Welsh onions (from the German *welsche*, meaning foreign) have been eaten in China and Japan since pre-historic times. Chives, shallots and leeks are all ancient and widespread in cultivation. And there are many more local varieties: the leek-like *kurrat* of Egypt, the coil-stemmed garlic rocambole common in Denmark, or the chive-type *ramps* of the Appalachians. I well remember the aura surrounding a friend of mine whose American Indian ancestry had led her to attend the Cherokee Festival of the Ramp in North Carolina. The Cherokee children were not allowed back to the local school for a week, the smell was so pervasive and persistent.

Those unaccustomed to using garlic in the kitchen are well advised to develop their taste gradually. A garlic press costs only a couple of pounds and will ensure that no large pieces are loose in the salad bowl. A pressed clove can be added safely to a small jar of oil and vinegar salad dressing, or to sautéed sauces for most casseroles or pasta dishes. Sautéed lightly in butter, pressed garlic can be dribbled into half-sliced French bread for garlic bread, or on to toast; or cubes of dried bread can be stirred into it to make croutons. Aioli is a simple garlic mayonnaise made by stirring the beaten yolks of two eggs into two large pressed cloves and adding olive oil drop by drop until it reaches a firm consistency: delicious on baked potatoes.

Colin Boswell enjoys garlic in his food as well as his fields. But he does not proselytize on the medicinal virtues of his harvest. The farm labourer who wore a clove of garlic inside his sock to ward off colds hardly fits the image of the supermarket sales drive. Nor does it fit the image of Höfels Pure Foods Limited of Bury St Edmunds, but their marketing director David Roser would be quick to defend the efficacy of the labourer's precaution.

In 1920 Dr Johan Höfels discovered a way to lock up the essential oils of garlic so as to avoid the odour and aftertaste. He called his capsules 'garlic pearles'. These are swallowed as a daily preventative tonic and as a specific remedy for ailments. "I took them all the time I was in Africa," David Roser said, "and they saw me through everything. I didn't even get dysentery." Höfels must be careful when making medical claims for its products, for it does not have the means to subject such claims to the battery of animal trials and large controlled studies now mandated for approved drugs. But others have been less circumspect. Garlic is lavishly praised in all the old herbals, and especially by the influential seventeenth-century Culpeper. Garlic has been used to combat both internal and external infection, and during the Second World War it was known as Russian penicillin.

Modern scientists have corroborated several of these claims in part: in the 1950s it was shown that oils of garlic consist of sulphides and disulphides which can unite with a virus and make it inactive. The Japanese scientist Fujiwara has proved the ability of garlic to increase the assimilation of B vitamins. It has also been shown that garlic oil effectively reduces serum cholesterol, and that a dose of between 2 and 6 mg. a day reduces the incidence of arteriosclerotic lesions in animals by one third. According to David Roser, at least 35 papers exploring the effectiveness of garlic have appeared in the medical journals over the last three years.

Medical herbalists often prescribe garlic today for circulatory disorders, for upper respiratory tract infections, and for infections of the stomach and upper and lower gut where, unlike conventional antibiotics, garlic does not destroy the benign bacteria which aid digestion.

Höfels grow their own garlic using organic procedures, and supplement this for their pearles with carefully vetted imports from Egypt. Fresh garlic may be more effective, but it can be less predictable: a herbal practitioner told me that she once prescribed a clove of garlic each day for a lady with persistent, severe catarrh who could find no relief using conventional treatments. This sceptical new patient returned three days later full of cheer and breathing clearly, but her mischance was obvious to all within yards of her, long before she could confirm it — she had innocently munched her way through an entire bulb, instead of one clove, each day.

Nor is the clove in the sock as silly as it seems. Put a garlic poultice anywhere on your skin and you will smell the garlic on your breath within two hours.

Colin Boswell does not grow his garlic according to organic principles. And yet he is a young man with foresight and he admits, or insists, that this must include a sense of responsibility for the environment. His father Martin thinks that the environmentalists sometimes go too far; that a particular "line" becomes fashionable and gets pushed to extremes. The older Boswell described his current role at Mersley Farms as that of a troubleshooter, though he was not talking now about the sometimes troublesome aspects of modern farming practices. "I try to prevent, by discussion, the worst excesses of youth." The issue he gave as an example was Colin's impetuous rush to install a computer to manage the payroll and accounts.

Martin Boswell took me on another tour through the same garlic farm. He showed me his collection of sea urchin fossils, pottery shards of the Beaker People, and some of his thousands of flints from prehistoric sites on the farm. Then he explained the detective work

which had led to his discovery of the Roman granary; showed me
Roman nails, the enormous bowl he was piecing together, and
photographs of the dig organized with the local archaeological
society. "There's a Roman villa in the parish of Brading, and one on
the other side of us at Arreton. But none here in the parish of
Newchurch. Not yet, that is!" Then we walked down across the silt in
the valley floor to the peat bog. "It's twenty-seven feet deep — that's
twice the depth of the Somerset Levels. But it's only economical to
work to eight feet because you have to pump so much water out.
We've got planning permission for 10 acres. It leaves a lovely clean
swimming-pool doesn't it? My wife swims two and a half lengths
every day." He had found bones at the edge of the peat and expects to
find hunting artefacts where the animals came down to water. "I've
seen what they've pulled out of Danish bogs, whole bodies with flesh
and skin intact." He showed me where they bagged the peat, for sale as
a soil conditioner. "We burn it on our fire, but there's no great
tradition of it because there's always been plenty of timber on the
island."

He told me that he had considered growing willows here, for
cricket bats.

"There is a tremendous esprit on the island, you know. It's like a big
village. You see a DL car registration and you know that they are
island people. You identify with them."

Garlic and cricket bats — and an identification with the ridiculous
notion that the one is as British as the other.

At the end of each season the best garlic bulbs are selected to be
broken into cloves to seed next year's crop, giving rise to a slight but
progressive difference between the KingCob harvest and the original
imported parent stock. So that eventually we may have an identifiable
British garlic for our new British taste.

11

Wielding the Cane — Raspberries in Blairgowrie

J. M. HODGE, SOLICITOR in the small Scottish town of Blairgowrie at the beginning of this century, was by all accounts an enterprising man. He took a keen interest in matters agricultural and in particular noted how well raspberries grew in the gardens of the cottagers who were his neighbours and his clients.

Raspberries grow wild throughout much of Europe and Asia. Our ancient ancestors did not know that the fruit contained useful amounts of vitamins A and C; niacin, thiamine and riboflavin; and the minerals calcium, phosphorus and iron. But they no doubt discovered at a very early stage that they were good to eat. As a rule of thumb, sweet things are safe to eat, and although the ancient raspberry was no rival to the selected varieties we know today, it must have compared very favourably with the sour crab apples and morello cherries, and the bitter sloes and bullace plums. Nurserymen began selection and improvements in these orchard fruits as long ago as Roman times, but with the exception of a reference to the wild berries by Pliny, the raspberry does not feature in ancient Greek, Roman or Chinese writings. Such improvements as were made with the raspberry occurred gradually over the centuries as a result of cultivation in cottage gardens, the original canes being selected from the wild, much as were the cottagers' little woodland strawberries.

With the building of the railways, fruit and vegetable production was no longer limited to market gardens close to the districts with large populations. Fresh crops could be grown on a larger scale in areas which particularly suited them, and still reach the markets quickly. In the case of the raspberry, however, the railways seem to have had a greater influence on the propagation of the fruit in the wild than on its cultivation in newly accessible fields — a July hike along many stretches of disused railways will soon reveal small raspberries fruiting along the unkempt banks, a testament to a widespread partiality for the fruit and the primitive waste disposal systems of the trains. For the raspberry is evidently *designed* to be eaten. It is both lure and reward for bird and mammal, whose droppings conveniently disperse the small, tough seeds. As far as the plant's *cultivation* was concerned, this remained a remarkably casual affair.

This was the situation in Blairgowrie as late as 1893, when J. M. Hodge, solicitor, observed the garden raspberry with great interest. The sugar tax had been repealed in 1874 and the jam industry was rapidly developing. Britain was destined to produce more jam and marmalade than all the other countries which would later form the European Economic Community put together. J. M. Hodge would dedicate his book *Raspberry Growing in Scotland* to Sir Wm P. Hartley, "A prince among preserve manufacturers, and a good man from whom I have received much kindness". And at one stage an estimated two-thirds of all the raspberries in the Common Market would be grown in Scotland: Blairgowrie would be the centre of a concentration of raspberry cultivation unequalled anywhere in the world.

George Hodge now occupies the same office in the centre of Blairgowrie as did his father, J. M. I tracked him down during the last week in July when the berries were at their peak, and he seemed to be spending more time with the growers in their fields than at his desk in town. The office was staffed by his son Andrew, a third generation raspberry solicitor.

Blairgowrie and Rattray, west and east of the river Ericht respectively, form a small township of some 8,000 people at the foot of the Scottish Highlands. About twenty miles northeast of Perth and northwest of Dundee, they lie on the northern edge of the Vale of Strathmore which is the "strath" or "broad" vale between the Highlands and the Ochil and Sidlaw hills. Fertile soils and a moderate climate encourage arable farming, with many acres down to oats, barley and wheat, and maincrop potatoes. There is also a large range of vegetables grown. The light, neutral soils which drain well have proved ideal for raspberries, as has the climate. Raspberries have a shallow, wide-spreading root system and need a certain moisture in summer, preferring the cooler north to the hot, drying suns of southern and eastern England.

"J. M. was no farmer," George Hodge explained. "He was a solicitor. Everyone called him 'J. M.'. But he was involved with farming through his work and he had a great interest in it. He served on various Agricultural Commissions and travelled to Canada and Denmark for them. He was especially interested in smallholdings.

And he was an enterprising man. He saw these raspberries growing so well in the cottagers' gardens and he saw a future in it.

"He persuaded one of his clients to plant out 2½ acres. Then in 1895 he started the Fruit Growers Association and marketed the fruit himself. In 1898 he took the lease of Wellbank. That was twelve acres and everyone said he was mad. They said he'd flood the market.

"In 1903 he formed a new Company and bought an estate at Essendy. It was 450 acres and he sold off 200 of them in 5–25 acre lots to individuals, to be planted out with raspberries. The Company brought in the pickers and built dorms for them. It was later called tin city and it was finally bulldozed just last year. In bad years the Company would carry over the payments, but the smallholders gradually paid off what they owed for the land. Everyone made money with the raspberries.

"Almost all the fruit went to the jammers, then. It would be packed with a layer of sugar on top, and sent on the passenger train. Then they used a sulphur dioxide solution and that made things much easier. When you heated them up the sulphur would evaporate. But that's not used much nowadays when everything is frozen.

"I farm 300 acres now and still grow raspberries. But the varieties have all changed. They deteriorated through virus infections. We've still got a few acres called the Antwerp field, after the berries we used to grow there. The Antwerps were one of the oldest of the named sorts. There were Antwerp Yellows, too. They were called white raspberries but they were a kind of golden yellow. And we used to grow Lloyd George. . . . My father was a great liberal."

Next to the solicitors' office, the Tourist Information Bureau was full of holidaymakers: "Blairgowrie's a bonny town. We get taken over by tourists once the holidays start." But the leaflet I requested was not on show. This was a list of local farmers in need of pickers. "We hand this out to the Glaswegians when they come for the raspberries. They don't want to know anything else. They just come in for this piece of paper and then they go, without saying a word."

Opposite the name of each farm were various comments: "Dormitory accommodation full" . . . "Casual pickers only. No camping facilities" . . . "Families and students preferred". One grower on the south-facing slopes near the town, advertised limited accommodation with showers, w.c.'s and cooking facilities. I walked up the hill to this typical raspberry smallholding which he has farmed for many years.

Beside the tidy rows of fruit was a small camp with three old caravans. Washing was strung up to dry between them. Children played barefoot among a heap of empty plastic baskets, and an adolescent had half of the engine of his Ford Cortina in pieces around him. The pickers moved from me quickly and one of the men waved me away aggressively. "Don't take any pictures, whatever you do," a younger woman insisted urgently, as if my personal safety depended on it. I hid my camera and tried to draw her into conversation.

"They're Glaswegians," she said, as if that explained everything. "And they won't be coming much longer. They want to tax us, that's why. The tax man. They want the farmer to write down all our names so they can take tax out, when it's slave labour as it is. If they take tax out, it won't be worth it any more. We get paid $7\frac{1}{2}$ pence a pound! It's terrible money! That's 45 pence a tray! I picked just over a hundredweight yesterday . . . that's a lot of picking . . . and I walked away with £9! The picking price hasn't gone up for six years. . . ."

The grower pays cash by the tray with no names and addresses, and that's the way he wants to continue. "They're clamping down on the 'black economy'," he said, "and they think we're part of it." The delicate raspberry, with its velvety texture and aroma, exudes a sense of luxury, and yet it is produced by a hardy plant which can survive in a tough environment. There is nothing delicate or luxurious about this grower and the raspberry operation which is his livelihood. He is a tall, wiry man. He wears blue jeans and heavy boots, with his shirt sleeves rolled up. He is coming to the end of a difficult season. "It's been too short, all the fruit coming at once. We had a dry, hot spell at the end of May so all the fruit set at the same time and now we've got this dry heat in July and with the two together all the berries are ripening at once. Three weeks' picking instead of six." He put his feet up on a cluttered desk, leaned back in his chair, and stroked the stubble on his chin. "We sell to everybody: canning, freezing, pulp, the fresh market. The pulp goes for jam, yoghurt, ice-cream. The best fruit goes for the basket, but they want the fresh fruit mostly at the end of the week, for the Friday markets, and anything over that goes to pulp. We sell the pulp in barrels — 50-kilo tubs — and they come and take them away. It just goes into a red mass. Folk often come in wanting to buy some raspberries for jam and I say we've got some in the tub and I show them and they say, 'We don't want that,' but I say it'll all be that way when it's made into jam. . . ." He knocked his cap back as he laughed and then pulled it down again with purpose. "You can hit high prices with the fresh fruit, in theory anyway. We're in some co-operatives, but we mostly send direct ourselves. For small lots we use the co-operatives but if we can get a full load we take it ourselves. I drive the truck — it does me good to go down a night — but I've never been further than Manchester."

The pace slackens after the picking, but there's a lot of work again when the leaves drop in November. The canes which have borne fruit will not bear again and must be cut out; and the new shoots which have grown up around the old this summer must be tied to the wires which run the length of the rows, to support them against the wind. He pulls a long sharp blade from a drawer and brandishes it in front of me. "That's what we use. I've started getting squads of men in. I pay them full-time. They slice the canes and lay them down and then the women come along to tie in the new, and I go along with the chopper. In the old days you carried the old canes away and burned them, to get rid of the diseases. That was really laborious, carrying the old wood out, but there's actually some good in the old canes. They've built up a kind of immunity. There's a liquid comes out of them when they're chopped up and it helps the new canes. So I'm told anyway. . . ."

New ideas in the cultivation of raspberries seem to emerge as frequently as new varieties, although in reality they represent years of work at the research stations at Invergowrie, near Dundee, or at East Malling, in Kent. Here there can be plantings of over 6,000 distinct seedlings, the researchers forever looking for raspberries with more fruiting laterals per cane, and more, larger and easier-to-pick berries per lateral; for canes with a higher resistance to disease; and for berries with a better sugar/acid ratio, a more aromatic flavour, and which exude less juice under pressure or when thawed after freezing. A novel way to approach the post-harvest labour, now under study, is to treat the crop as a biennial. The young canes in one half of a field are burned down by a chemical as they emerge in early summer so that the fruiting canes can bear their crop without the shading and competition of the emerging new canes. In the following year, these rows will produce an abundance of young canes without the competitive influence of fruiting canes, and so a cycle of each half of the field cropping every two years is established. In the 'on' year the crop is heavier and much easier to pick than normal, theoretically providing 85 per cent of a normal crop, with much lower labour costs.

Nairn Thomson is not convinced by early results of a small trial of this particular method on one of his farms. But he grows his 100 acres of raspberries with a keen eye on the future. In his suit and tie at his smart office at Keathbank Works, down by the river, he is a modern businessman in appearance and in practice. He watches the raspberry acreage in Kent increase each year, and above all he watches the

imports of cheap pulp from Eastern Europe. He appears to want to
listen as much as to talk, although he is eager enough to extol the
particular virtues of the Scottish raspberry.

"We were involved in textiles as well as farming. There were fifteen
mills here on the river, based on flax for linen, and then jute, and then
man-made fibres. We closed the last of our mills in 1980. It's one of my
theories that the raspberries were made possible here because we had
this population, because we had these pickers in Blairgowrie. You
need a good supply of temporary labour." Then he described for me
the rest of the work on a raspberry farm, apart from the picking: there
are usually around 3,000 plants per acre, and these must be replaced
when the stock becomes exhausted after ten to fifteen years. The canes
are planted two feet apart in rows with six to seven feet between them.
They need this space to receive enough light, so that they can fruit all
the way up the cane, but it also allows access for large machinery.
Herbicides are applied to control the weeds. In June insecticides are
sprayed on against the raspberry beetle, and fungicides against mould.
And the crop is irrigated during dry summers.

One of the farms which Nairn Thomson now operates was among
the first to be rented out to J. M. Hodge. "He was the first, but others
followed soon. Robertsons in Paisley were the biggest jam people I
remember, and the raspberries went by train. But it was a terrible
thing when a truck would get lost in a siding, and it happened all too
often."

Now the "terrible thing" is the cheap, low-grade frozen pulp which
is dumped on the British market "all too often" by Poland, Hungary
and Yugoslavia. Sales of fresh fruit to the major wholesale markets are
an important and profitable business, but 95 per cent of Scottish
raspberries, which are now grown on some 6,000 acres throughout
Strathmore and the Carse o'Gowrie, still go for processing. Manufacturers will pay a premium for Scottish raspberries, and they have a
high-quality export market of their own, but the price of low-grade
foreign pulp inevitably influences the market price.

According to Murray Cormack of the Scottish Crop Research
Institute, the Scottish growers can only compete because their picking
costs are low. "It is easier to pick for pulp than for the basket. Kent
growers obain much higher returns on the fresh market but have to
pay nearly three times as much per pound picked."

The Institute has bought a raspberry-harvesting machine from its
Oregon manufacturers and are looking at the biological aspects of its
commercial use in this country. The machine shakes the canes with
flexible mechanical fingers, and the ripe berries fall on to a belt below.

"It could be used with current varieties," says Cormack, "if the labour situation warranted it. But meanwhile we can look for more suitable varieties, which ripen uniformly and stand the treatment better." And then he launches into a pleasant and optimistic account which seems at once like a PR job for the raspberry, for his Institute, and for Scotland itself. "Remember, these things can take decades to develop," he says, with a cheerfulness as bright as the berries themselves.

"Actually, we see ourselves as working for the housewife. The farmer's just the middle man. We're working for varieties with higher yields and resistance to disease, which means cheaper fruit; for higher quality, more palatable berries with canning, freezing or fresh handling suitabilities. Our Rubus breeder, Derek Jennings, has bred the Glen series of raspberries. Over half the acreage in Scotland is Glen Clova now. Glen Moy is a new one for the gardener, a lovely fresh fruit. Then there is the River series." These are hybrids of raspberries and blackberries, in the manner of the loganberry, first grown by Judge Logan of California, and the American boysenberry. "Our tayberry is perfect for the Pick-Your-Own trade and is grown all over the world. It's even larger than the loganberry — a deep burgundy colour with a rich aromatic flavour. It makes a lovely jam. And now we've got a new improvement in the tummelberry." The raspberry derived its botanical name *Rubus idaeus* from Mount Ida in Asia Minor, on whose slopes it flourished centuries ago. The genus rubus includes the blackberry, the Japanese wineberry, the American dewberry, the cloudberry which is known to fruit naturally within the arctic circle, and many others. There is a diverse store of genetic material to work with.

At the East Malling research station in Kent, new varieties such as Leo and Joy have been bred for the lucrative late market in the south. Raspberries emerge from the freezer tasting much more like the fresh fruit than do strawberries, but they do tend to slump and bleed as they thaw. Apparently, if you sit some frozen Glen Clova berries two inches apart on a standard filter paper their juice will spread all over it, while Leos leave spots hardly bigger than the berries themselves. But I gained the impression that the Scottish growers are not too concerned about the exotic market — "autumn fruit for London hotels and so on" — and are quite content with their native varieties.

On my way back from Blairgowrie I stopped at a Lancashire wholesale market. Raspberries grow satisfactorily on most garden-soils and are an essential feature of most Pick-Your-Own operations, and yet there is invariably a strong demand for the fruit on the traditional market. All the salesmen had cleared their lots. One of them bemoaned the loss of the railway. Two young lads had driven a van down from Blairgowrie "and bounced the fruit something wicked". Arriving in the early hours they had caught a nap before trading began and the still-ripening berries, sitting unventilated for just three hours, had "boiled" in their own heat. They had followed their drivers' example and "gone to sleep". This, with its tell-tale stain on the bottom of the punnet, is as familiar a habit to those in the trade as "growing fur coats" — which raspberries can also accomplish in a matter of hours if picked in the rain. But the salesman had still cleared the load at a respectable price.

Perhaps these "sleepy" berries made their way into jams, jellies, tarts, flans, pies and sauces — or into the traditional British summer pudding, which is surely a fitting destination for even the best that Blairgowrie has to offer. In Lancashire they line a large pudding basin with $\frac{1}{2}$-inch slices of stale bread. Then they pour in about a pound of stewed and sweetened raspberries, cover with more slices of bread, and then with a saucer and a heavy weight. A flat iron was much touted for this role in the past. After four hours the pudding can be tipped out, upside down, and dressed with cream or custard. Simple and sensational.

There is a problem, nevertheless, with the raspberry dessert. That sweet foods were mainly safe may have been sound guidance for our prehistoric ancestors, but hardly applies in an era of sweetened processed foods when each Briton consumes an average 130lb. of refined sugar every year. Quite apart from the more serious degenerative diseases, this is the era of dentures, and as I learned from my own immediate ancestors, the dental plate and that tough little raspberry seed do not get on well together. Or rather they get on too well together, the seeds lodging underneath with a speed and certainty unrivalled by any other commodity, and rendering the continued enjoyment even of summer pudding quite impossible. Of course, the researchers at the raspberry institutes are working on producing a seedless berry. . . .

12

Summer Squash on the Riviera — A New
Crop for Cornwall

———————————

THERE ARE PLENTY of farmers in Cornwall who still talk about sending their produce away to England.

Dick Jelbert comes from a family which has lived on the Penwith peninsula for many generations. The Jelberts were wheelwrights originally, and they were one of the four original families on the Parish Register when Gulval Church began its records. But Dick has been a farmer all his working life, and his son Michael now works with him. They both speak with the quick, singing quality and gentleness of the Cornish accent, which was bequeathed them by the Cornish language of their ancestors. This language went into a rapid decline under the influence of the social revolution in the Tudor period and with the coming of a new trading and maritime importance for the far west. The mineral resources of the peninsula became increasingly important to the rest of the nation with the development of industry, and the building of the railways finally ended the isolation. The climate of the 'English Riviera' between Penzance and Helston drew the English down to the southwest, and the early crops which that climate could produce could be carried swiftly away to English markets. France, Italy and Spain became the Cornish farmer's rivals, and the intricacies of international marketing their concern. As I ventured for the first time to visit growers who had always been for me a name on an invoice or a voice on the end of the telephone, I was struck by a particularly sobering symbol of the irrevocable end of Cornwall's isolation: the seaweed once gathered by her coastal farmers as a crucial aid in improving the generally poor soil was reported to be contaminated with radioactive waste from the distant coastal power stations of France and Britain. John Wesley had preached in the village where the Jelberts live. Now anti-nuclear campaigner Dora Russell, from her house at Land's End, said that we need another Wesley, to stand at every village corner and shout out, as he did, "Repent ye, for the end is at hand."

Dick and Michael Jelbert describe the disease which struck their anemone crop in 1976 and the frost which devastated their early potatoes in 1982 as if they had experienced an uncomfortably close vision of their own particular end at hand. But farmers on the productive slopes around the English Riviera have been searching for many years for new crops with which to supplement the spring flowers, broccoli, spring cabbage and early potatoes on which their prosperity has depended ever since the coming of the railway.

The Jelberts work 17 acres on the slopes of Bone Mill Valley near Heamoor, just a few miles above Penzance. In the patchwork of fields on the south-facing slopes between here and Helston, around the curved instep at the very foot of Britain which is Mounts Bay, a remarkable crop of cauliflowers is produced. In a good year some 20 million head are harvested from this small strip of land, in fields whose average size is less than 3 acres. Planted in spring, the cauliflowers are not harvested until the following winter or spring, by which time they have developed a far superior taste and texture to the fast-growing summer varieties which are sown and harvested in the same season. Enjoying the mild climate afforded by the warm Gulf Stream, and the somewhat longer daylight hours (unreduced by fog) of its southerly latitude; protected from occasional frost by its envelope of tightly wrapped leaves, and from the wind by the frequent walls and hedges around its small fields — this crop reaches a winter market eager for variety in its limited supply of vegetables.

Locally this vegetable is still called broccoli, by which name winter cauliflower was once widely distinguished from the quick-growing summer varieties, although with the increasing popularity of purple sprouting broccoli and summer calabrese, which is also sometimes called broccoli, this particular usage is now limited to Cornish growers and those in the fruit trade. There is a local legend that good broccoli can only be grown in view of Mounts Bay, and growers there defend its quality passionately. "You've got to realize that broccoli starts to die as soon as you put the knife to it. Our broccoli is likely to have been cut in the last twenty-four hours. We can get it to many markets in less than twelve. But this other stuff" — by which they mean French, Italian and Jersey — "is days old by the time it reaches the plate. You can tell the difference, because ours is full of sap and growth."

Until 1945 all Cornish broccoli was carried away by rail. In 1931, for example, "there were four or five goods trains every Monday and Thursday during peak periods, with no fewer than seven specials — 300 trucks — leaving West Cornwall one day for London." Now the

Jelberts sell their broccoli through commission agents who collect it from the farm in their own lorries. As soon as the broccoli is cleared in mid-February and early March, the early potatoes are planted. They are busy through March and April with the violets and anemones which they grow for cutting and bunching and which in a good year return their best profit. Then in late May and early June the Home Guards are ready for lifting, hopefully to fetch a good price before any of the home-grown rivals from Pembroke and Kent reach the market.

"In the old days," Michael Jelbert said, "you didn't have to fill the summer gap. You could afford not to, with a good broccoli crop behind you. But now you've got to work all year round, you've got to fill every little gap there is. So then the question is, what with? In the past we've grown peas and carrots and beans and so on for the local trade. A lot of tourists come down here in the summer. But that's nothing of a job. We've got a few broad beans in now, and we can't give them away. So last year we asked the man from ADAS for advice and he put us in touch with Gloucester Marketing Services (GMS), which is a big marketing co-operative with headquarters in Cheltenham . . . they have a packing house a few miles from here at St Erth, where they pack for Marks and Spencer and so on. They suggested we grow courgettes, which we'd never done before. But we got advice from ADAS and we went ahead with just a ½-acre or so and it worked out all right, so we're doing a bit more this year, with some under polythene tunnels as well. The idea is that we get the early market, through GMS, before anyone else is ready. We keep that market with regular supplies through the glut when everyone can grow them all over the country, and then we go on later than everyone else at the end of the summer when prices pick up again, because of our climate.

"I think there are about 15 of us growing courgettes for GMS, mostly on about 1 acre each. They like small-scale growers because we use family labour. Otherwise it gets too expensive.

"We plant the seeds in mid-March and then plant them out three feet apart as soon as they've grown two leaves. We're using a variety called Ambassador. We plough in manure first because they're very heavy feeders. There is still a dairy farming business down here so we can get the manure. Then we put a bit of blood, fish and bonemeal on each plant once they're in. After they are properly established we use nitrogen pellets. We spray for aphids and blackfly, and we put on a mildew spray as a precaution. And we have to protect against slugs or they'll do terrible damage. We pump water up from the river with the tractor, because they need plenty of water. And that grows lots of weeds. We have to hoe the weeds by hand because we'd do too much damage to the courgettes if we sprayed. But the hardest work of all is in the picking. The plants in the polythene tunnels are ready for picking in just six or seven weeks, and the ones outside by mid-June. Once we start, we pick until October, and we have to pick every day. We can only sell them up to eight inches long — six inches is ideal — and they grow so fast you don't have time to turn round. So we pick every day, first thing. It's hard work. We wear these waterproof leggings or else we'd be sodden wet from the dew, and these plastic sleeves or we'd get our arms cut to pieces on the prickles."

I left Michael and his father picking through the tunnels and walked up the steep slope, where the outside courgettes had grown just sufficiently close to ensure that my trousers became drenched. The rising sun caught the bright orange and yellow flowers which rose everywhere between the broad leaves of the plants, and I thought that the fields of anemones and violets which lay to either side of me, at this stage full of tiny seedlings, could never look any prettier. This was agriculture, as opposed to market gardening, but it was on a very human scale. It was perhaps ironic to hear the work described as 'grand' as well as 'hard'. Just as it seemed ironic to me on these sun-kissed southern slopes that, with the exception of these courgettes, the great advantage of the sun and light was used to grow *winter* crops. And yet both appeared eminently reasonable.

At the top of the field, looking back down towards Mounts Bay, I knelt to feel the soil. It was a lovely sandy loam, light and warm. Originally, I had read, this soil was a heavy clay. Further east in the Tamar Valley, famous for its early soft fruits, the smallholders of the last century had pulled up soil from the valley bottom using pulleys and sledges, to clothe the thin slopes. Around Mounts Bay they composted the haulm from the fields with layers of sand and seaweed. "There were always farmers down on the beach, loading seaweed on to carts. There were enormous heaps alongside the fields. . . . There must have been three or four feet of soil added."

When I called later that day on the Rosewarne Experimental Horticulture Station, a spokesman said that he supposed the quality of the soil in general would be gradually deteriorating now. Although many farmers, like the Jelberts, were able to incorporate some manure, there was far less organic matter going into the soil than formerly. And there was a problem of soil erosion through occasional freak storms with very heavy sudden rainfall, to which the area was prone. But he thought the prospects for the Mounts Bay area were very good. "The anemone business is holding its own, and there has even been a resurgence as far as broccoli is concerned in recent years.

New strains are evolving which ensure a much higher quality, so that 80 per cent of the crop can now be sent to market instead of the 50 per cent of the past. And a decline in the early potato crop has been more than offset by the potatoes grown under contract for crisping. These are the main crop, Record, but we can pull them earlier down here, so the crisping factories can get to work earlier, and it is proving a good proposition. Then there are plenty of trials with early crops grown in polythene tunnels, such as carrots, calabrese and courgettes. Courgettes seem quite attractive."

If courgettes seem attractive to more than a moderate number of farmers, however, there is a danger that the market will very quickly become saturated. This is not because the demand is static — in fact it is growing considerably each year with more than 6,000 tons grown in the United Kingdom — but because the plants are so prolific and the yields so high from such small areas. This is common, and happy, knowledge to many beginner gardeners for whom courgettes serve much the same role as do radishes for children: they germinate readily and crop early, providing such bounty that proof of success can be offered generously to all relatives and neighbours. To the more experienced, the yields quickly become an embarrassment. A small fortune, according to the folklore of the suburban gardener, awaits the first geneticist to breed a half-a-courgette plant.

There is an almost indecent fecundity across the entire range of the family Cucurbitaceae, including cucumbers, melons and gourds as well as the 'squash' group to which courgettes belong. There is a variety of Far-Eastern cucumber whose flowers open at dusk and close at dawn and which soon after pollination grows at a rate of two inches per day, to a length of five feet. The waxgourd vegetable of the Asian tropics is protected from micro-organisms by its waxy coating, allowing it to be stored for up to a year without refrigeration. At one stage the plant grows nearly half-an-inch every hour, a rate which allows three or four crops to be grown each year. The Mexican buffalo gourd, whose seeds are high in protein and oil, can produce up to 200 fruits from each plant. The roots can reach fifteen feet deep in search of water and keep the plant alive for decades. And the chayote of Central America contains only one seed, but with the unusual property of germinating while still in the fruit. Closer to home, we have had a Lincolnshire farmer growing 300 marrows, weighing almost one ton, from a sixpenny packet of seeds. And our prize-winning pumpkins, making new records almost every year, are poised to break the 500 lb. barrier. In China the pumpkin is called the Emperor of the Garden, and it is the symbol of fruitfulness.

A tremendous range of varieties is also a feature of the family, and of the particular species, *Cucurbita pepo*, to which the courgette belongs. Members of this species have been called 'vegetable marrows' in Britain, but the American term 'squash' is becoming more frequently used. There is considerable confusion regarding classification in both popular and scientific terminology, but I suspect that with increasing exposure to a broader range of squashes we shall soon follow the Americans' example and divide them from the consumer's point of view, into the summer and winter types.

Unlike the cucumbers and gourds which originated in the Far East, and the melons from central Asia, the squashes are native only to the Americas. Archaeological digs in Mexico have yielded the seeds of cultivated varieties dating from 4000 to 9000 BC, which would indicate that squash was the first food to be cultivated in the western hemisphere. Squashes cross-pollinate very readily, so that over the millenia a great diversity of form and colour has developed.

The summer squashes all grow quickly and are picked in an immature condition, when the skin and seeds are still soft and edible. They must be eaten quickly as they deteriorate very rapidly once picked. Almost all the summer squashes — long, round or scalloped — are available in white, yellow or green, striped or mottled, smooth or warty. In the USA they have colloquial names which seem to change from county to county. Amongst the most common are the yellow Straightneck, orange-warty Crookneck and Pattypan or Custard Marrow which looks like a white flying saucer. The winter squashes, on the other hand, are mature fruits which have grown slowly and are harvested late. They usually have a tough skin which is not eaten and large firm seeds which are scooped out before cooking. The skin protects the pulp inside so that they can be stored for three or four months before eating. They include the pumpkins; the knobbly Hubbard; the dark green, round Acorn, with slight longitudinal grooves; Butternut which is a dumb-bell shape with a smooth tan skin; and Spaghetti Squash, whose flesh when cooked and scooped out comes away in long thin strands like spaghetti.

Most summer squashes are composed of about 92 per cent of water, are quite bland in flavour, and are usually cooked with cheese, onions or tomatoes. Whereas winter squashes tend to be tastier, sweeter and more nutritious, with a denser and usually more coloured flesh. They can be baked and filled with various stuffings but are often served with simple seasoning as a side vegetable.

The courgette is the most popular of the summer squashes, the word being derived from the French 'courge' meaning 'squash' in general. There is a very similar plant with the Italian name, zucchini, which is widely grown in the USA. They were both developed from selected strains of marrows to provide succulent small fruits. If left on their vine, however, they will grow and mature to become indistinguishable from the large tough-skinned marrows which are often more welcome as the harvest festival centrepiece than as a food at the table. Some of the other summer squashes, if left on the vine, will become winter squashes, but other types are only suitable for eating when mature.

Although some of the squashes were brought to Europe soon after Columbus, they have never become very popular. Dorothy Hartley was obliged to borrow American recipes for pumpkin pie and soup for her classic book *Food in England* and admitted that there were no traditional recipes for marrows as the vegetable only came to us at the end of the last century. The courgette has been here only since about 1950. But we now spend some £2 million on the vegetable each year, and British growers provide all our main season supplies.

Back down the slope, Dick and Michael Jelbert had finished picking and were weighing their crop into 12-lb. cardboard trays for delivery to GMS. It was breakfast time and the day was warming up. I asked the father if he ever ate any of his courgettes and he answered "Yes" a little over-eagerly. "Yes, they're all right if you fry lots of onions with them." I told him that the Iroquois made a feast out of the flowers and he looked back over the field suspiciously. I suspected that this was more of his son's project. And what about all the other squashes? I told them that I had seen acorn, butternut and spaghetti squash on the shelves of my local supermarket, imported from the Mediterranean countries and selling at £1 each. Were they tempted to branch out? Could they be the first into new markets?

The father seemed terrified. He looked back up the field as though at the devil he knew. The son was simply practical. "There's only any point in growing what we can sell, and GMS only want courgettes. We can't supply your supermarket up in Shropshire. And when these courgettes drop to 10 pence a pound in the next few weeks, we'll be wondering enough about *them*."

13

One for the Birds — The Kent Cherry

————————————————

I FOUND PERCY NYE up a twenty-foot ladder in a cherry orchard near Teynham, Kent. It was the loveliest of days in mid-July, with an embracing sun and clear blue sky. On the ground the air was close with the high-summer scents of hedgerows and grasses, while in the uppermost branches the leaves stirred slightly in a gentle breeze. As for the rest of the world, it seemed to be full of cherries. Deep red globes of fruit in clusters like galaxies consumed the field of vision, for the boughs on these 'loveliest of trees' were hung with one of the heaviest crops for years. Percy pulled a high branch down towards him and tumbled the cherries with nimble fingers down into his 'kibsie' or bucket.

"I must have been about twelve when I first had a go picking cherries, and I'm seventy-two now. I've lived here all my life, so that's sixty years of it, and the cherries are as good as ever. Some of the early Rivers, for instance, haven't altered since then. But plenty of things *have* changed. There were hundreds and hundreds of acres of cherries in this area when I started, but over the years it's kept getting smaller and I remember five years ago saying that in ten years there wouldn't be any cherries any more. There will only be a few trees head-high with nets over them, that's the economics of it. In the old days we used to talk about going up sixties . . . that's a sixty-stave ladder . . . sixty rungs at nine inches apart, you work that out. The trees would be seventy feet high then, and we'd stick sixties up and pull down the top branches to pick them. But you couldn't get people to go up a tree like that now. On a big tree in a good year you'd have your ladder in one setting all morning, you could pick from one position for hours, and you'd gradually work your way round. Then you'd put a little ladder like this one I'm on now up the middle. I remember the time we picked 1,000 trays on a Sunday, and they were big trays too, twice as big as these we use today. People came on bicycles and on foot, as well as in cars. Anyone that moved we had up those trees. That was an old variety called Governor Wood, a pale yellow cherry with a soft red blush, for canning. But nobody wants them any more."

Disgraces are like cherries — one draws another, as the old proverb says, and the old big trees have suffered one disgrace after another. They are most discriminating as to climate and soil conditions and the majority of our English summers fail them. The flowers are highly susceptible to frost damage, and untimely rain causes split fruit. They are plagued with more diseases than most fruit trees; bacterial canker will kill the tree and viral infections greatly reduce its vigour. The canning industry has deserted them. Pickers won't pick them, or can't pick them profitably. And now the latest and most devastating insult, from the birds. . . .

"Birds were here in the Thirties just the same, granted, but it wasn't a problem then because there were so many for them to go at. I mean there were cherries spread out everywhere. But now they congregate in just a few places. About five years ago we had so many starlings coming in we nearly packed up. They came over in clouds so you couldn't hardly see the sky through them . . . crapping all over everything . . . they'd come right down on you and you had a job to get them out. There might be seven or eight guns blazing away and they'd just come right down on you. . . .

"It hasn't been so bad this year. But cherries will never come back to the way they used to be. If it's not one thing, it's another. It's stacked up against them. I mean, here we are in the heart of the cherry district and the wife went out last week and they were selling American cherries at £1.30 a pound, when there's fruit like this that can't be beat, right on their doorsteps! It's bloody ridiculous. It doesn't make sense to me. I suppose it's all . . . I don't know whether there's backhanders in it or what. I don't know. It doesn't make sense to me, when they could sell these at half the price, but I suppose they know what they want to do. . . ."

Two species of cherries are native to Britain. *Prunus avium* gives us the wild Mazzard which is a splendid seventy-five-foot-high tree with a smooth silvery bark, white blossom, brown-tinted young leaves, and sweet, or at least non-acid cherries. *Prunus cerasus* is a smaller tree and its fruit is sour. This species gives us the dark-coloured Morello and the red sour cherries such as Flemish and Kentish Red, sometimes called 'Amarelles' or 'Griottes'. Botanists believe that the Dukes or Royal Cherries, which are hybrids between the two parent species, have resulted from the ancient domestication of our native cherries. But the rest of our older orchard varieties were introduced by the

Romans. These superior cultivars were derived from the two wild species in Asia Minor, and discovered there by the Roman General Lucullus during his campaign against King Mithridates. This soldier was also a wealthy epicure, with a keen palate for the best selections. The cherries had been greatly improved by a Scythian people who pulped the fruit to make a thick juice drink and a sediment which was then pressed into cakes and became part of their staple diet. From Rome these improved varieties were distributed quite rapidly throughout the empire.

Little is known of the early development of fruit orchards in Britain, although it seems likely that astute monks would select the better cherry trees in their monastic gardens, and thereby gradually develop some new local varieties. Cherries were certainly sold on the streets of London in the thirteenth century. By the early Tudor period, however, there was a general scarcity of fruit, especially near London, so that Dutch and French merchants sent regular consignments to be sold at markets such as Billingsgate. In 1533 one Richard Harris, with the title Fruiterer to King Henry VIII, introduced superior varieties of cherries, apples and pears from Flanders and planted large orchards in Kent, of cherries in particular, near Teynham and Sittingbourne. A Kent orchard of 32 acres is reported to have produced cherries in 1540 which sold for £1,000. This is an incredible return, considering that good land was then let at one shilling per acre. The fruit from these new trees could be transported easily by barge, round the Isle of Sheppey and the Medway estuary and up the Thames to the capital. Though the profits no doubt fell as more trees matured, the enterprise was so successful that by 1604 one writer reported that, "by reason of the great increase that now is growing in divers parts of this land of such fine fruit, there is no need of any foreign fruit, but we are able to serve other places".

The usual custom in Kent was to plant hops, filbert nuts, apples and cherries uniformly through the same orchard. The hops would be cleared out after about twelve years and the filberts grubbed after thirty years, by which time the cherries and apples had fully matured and required all the land. The larger-growing varieties of cherries were planted at least thirty-six feet apart. This system persisted on some farms until the end of the last century. Cherries were also grown in east Kent between Canterbury and Deal, on the Ragstone Ridge south of Maidstone, on the High Weald between Tunbridge Wells and Tenterden, in Berkshire and Buckinghamshire, and in the Teme Valley area between Ledbury, Worcester and Bewdley. But the area of north Kent around Sittingbourne has always far surpassed all others in importance.

Sometimes the grower sold his cherries on the tree to the highest bidder at auction; sometimes 'travellers' were hired to pick at piece-rates, and the grower would carry his crop to market in pots, or 'sides', containing 63 lb. of fruit. A writer in 1798 recorded a particularly plentiful year with great quantities of cherries but still a heavy demand. In 1838 the government replaced the large 3s.-per-bushel duty on imported fruit with a charge of 5 per cent, and large shipments soon arrived from France and Holland. But British cherries were evidently still profitable. Another writer recalls the excellent crop of 1886, when prices dropped from 14s. to 5s. per side in a single day, but demand held firm. Buyers came to the orchards on long, low carts which each held about 20 sides of cherries and, on one particular day, 200 such carts attended one farmers' market. By 1904 one authority concluded that where soil and situation were favourable, few hardy fruits were capable of giving, on average, a better cash return than the cherry. The acreage down to cherries continued to increase, from 11,000 before the First World War to 15,000 by 1935. In that year a Kent grower told an investigator: "Anywhere was good enough to plant cherry-trees in the old days. The fields had to be kept for corn! That was the paying thing. But I've had better crops and more money from those old trees than from almost anything on the farm. . . . People seem to be taking to them, the nice big red and black ones, for dessert. . . . We pack them into 12-lb. chips instead of the old pots. . . . It doesn't pay now to grow cheap stuff."

Today cherries are even packed in 1-lb. punnets; the big deep red to black ones for dessert, as opposed to the 'whites' formerly much used for processing, are now all the public appears to want. Both of these are sweet cherries, originally derived from *Prunus avium*. The 'whites' are in fact a pale yellow more or less covered with a red flush, the main one still appearing in the market from time to time being Napoleon. Another distinction was formerly made between the 'Geans' or 'Guignes' with soft juicy fruit and the 'Bigarreaus' or 'Hearts' with a firm, crisp flesh and 'crack' to the tooth. Now there are many intermediate forms, though most of the successful Merton varieties raised at the John Innes Institute tend to be black Bigarreaus.

As well as the size of the containers, the acreage has come down drastically — from a peak of 18,100 acres in 1951 to less than 3,000 acres, almost all of which are in Kent. It may well continue to shrink until we have only a quarter of this area.

Perhaps, despite the experience of centuries, the report of the Ministry of Agriculture was correct when in 1968 it concluded that the climate of this country, even in Kent, cannot be considered good for cherry-growing. Perhaps the fruit and its processed forms does not appeal to the national palate: the Germans grow half of the total EEC crop (if we include sour cherries) and consume a further 30,000 tons a year imported from their neighbours. We British eat only 12,000 tons of cherries annually, and less than half of these are home-grown.

And yet a merchant put a strong case for the British cherry crop in the trade press, at the height of the grubbing-up of orchards: "As a wholesaler handling cherries from a number of countries, I am saddened to think that there are many people today who have never ever tasted an English cherry, which at its best can compare with any that the world has to offer."

The farm at which Percy Nye worked for the last thirty-two years, and at which he still picks cherries in his retirement, is the Newlands Farm of the Boucher family. Potatoes, wheat and barley are grown on half of the 700 acres, and fruit on the rest. They have 120 acres of dessert apples, 12 acres of Bramleys and 30 acres of pears; 80 acres of blackcurrants grown under contract to Beechams for Ribena; 27 acres of strawberries and 8 acres of blackberries which are sold through the Kentish Gardens marketing co-operative; 12 acres of gooseberries sold on the wholesale markets and direct to supermarkets; and 36 acres of cherries.

Rex Boucher is currently Vice-Chairman of the Kingdom scheme set up by the Apple and Pear Development Council and subsequently adopted by Food from Britain as the Quality Food Mark for apples and pears. He spoke to me in his executive office which is part of a complex including packing houses and cold stores for 1,000 tons of fruit.

"The farm staff consists of 16 fulltime men, and we have a permanent staff of about 20 women, who are mostly employed in the apple-packing section in the wintertime. During the summer they are joined by up to 300 pickers.

"We used to have 250 acres of cherries, not so long ago. But now we're down to 36. We may end up keeping 10 to 12 acres of new, small trees under netting. The sheer excellence of good cherries means that a few specialized orchards will be profitable. There is a great demand for good fruit. But we've got to get the right varieties on the right root stock. The old cherries were forest trees, grafted on wild cherry stock. We've got this new root stock, Colt, and so on. But these new root-stocks were not as resistant to bacterial problems as we hoped. And it takes ten years for a new orchard to start bearing at full potential. So we'll be watching the trials, in fact we're running some ourselves for the East Malling Research Station. There has been a lot of very promising work on new varieties, started twenty years ago by Peter Matthews at the John Innes Research Institute. They've given us Merchant and Mermat and others. When we get the right variety on the right stock, then we'll come back with a specialized orchard."

Rex Boucher's son Hugh, who has recently stepped into Percy Nye's shoes as farm manager, then took me back out to the orchards. He picked clusters of cherries — Stellas and Nobles — as we walked and was constantly popping them in his mouth and spitting out the stones. "If you're not careful, you have the runs for six weeks," he laughed. He explained that the stables housed an equestrian and racehorse stud, which is his grandfather's interest. "Grandfather was a seed salesman who went out to Australia with his brother to earn his fortune. After five years there was a terrible drought and he packed up and came home. He bought a few acres here and built it up from that."

The ground was littered with empty cartridges from shotguns and rope bangers set to scare away birds. "They cost us about £2,000 a year," Hugh said, as he kicked one away absentmindedly. "This orchard is about 15 acres, and it's over forty years old. It's got about five years to go. The trees are just too big. You can halve your picking costs on smaller trees.

"We've got the right soil for cherries. It's a deep rich alluvial soil, sixteen to eighteen feet of brick-earth on top of chalk. It drains easily, and yet it retains moisture quite well in the summer. And we're relatively free from frost at blossom time. But the economics of cherries are very unreliable. This is obviously a very good year. You might have a high profit one year in five. You just creep along for two or three. The rest is disaster." Then he described the potential advantages of the new specialized orchards. "You can train the branches so that all the fruit is at arm's length. Unfortunately Colt stock is not as dwarfy as expected, but that will be improved in time. We've got to get much more intensive. The new German selections, for instance, will fruit after only four years. We can plant close, and to suit the machinery. We operate a scorched earth policy at first, with herbicides, and mulch with straw. Then later we can mow between rows. We can spray for insects pre-blossom and then with a bactericide after picking. And with trained branches, eventually we can put the whole orchard under net."

Hugh directed me to another orchard, where larger gangs of pickers were at work, and as I crossed the main road I pondered yet another factor in the decline of the British cherry. Throughout Britain approximately 15,000 acres of agricultural land are lost to urban and industrial development every year. How much of this relatively small strip of land in north Kent which is best suited to the cherry has been lost to the M2, which runs right through the middle of it, and to the new housing encouraged by ever quicker transport to the capital?

I was drawn under boughs hanging low with ripe fruit to the cheerful noise of lunch. Half-a-dozen toddlers were running in and out amongst the sets of scales and wooden trays, clutching peanut-butter sandwiches and packets of crisps. Their mothers were grouped together, seated on upturned empty boxes, each within her own circle of plastic bags and containers, and cans of pop, and all within a larger circle formed by each picker's stack of weighed and graded 12 lb trays of cherries. They worked from nine until three, with half-an-hour for lunch. On Saturdays and Sundays, freed from the school-time routines of older children, they picked from seven until twelve. And they appeared to love it.

"This is the best job of all," said Pam, who was the most instantly talkative. "I've done strawberries and gooseberries, and then in winter we pack the apples and pears as they come out of cold store. But this is the best job of all . . . in the fresh air, and the sun . . . with a bit of a breeze . . . and vertical! Oo, it's lovely, picking cherries is!" Then they all had their piece to say: "You get scratched a lot on gooseberries, and they give you belly-ache if you eat too many. . . ." "Cherrying is the best time, because you're out in the air with your back straight and there's nothing better. . . ." "It's the best money too. He's a fair man, really, Mr Boucher. He always finds you a job. . . ." "We'll show you the best trees, I mean for the best cherries. You get quite choosy after a while." The best two baskets of cherries at the Kent County Show are traditionally sent to the Queen and the Queen Mother, but I suspect that Pam, Kath, Peggy and friends fare quite as well as Buckingham Palace in this respect. "We're not supposed to take any home. . . ." "'Course, we're not saying we do, like. . . ."

Maurice called an end to lunch and the women grabbed their buckets and climbed up into the trees, several of them disappearing from sight but not from sound, as they constantly joked and bantered with each other. Maurice was the foreman or 'ganger', one of the regular farm staff, and he was at the centre of much of the tomfoolery.

"I've been a ladder man for fifteen years. I put the ladder in and then I go up it. I won't send them up without I go up first, and I wouldn't put them nowhere I wouldn't go myself. You can't lay it flat, or you'll break the branch with the weight. So it's got to be as vertical as you can make it. There's some trees, some sides, you just can't get a ladder in safe no way. You've just got to leave them. But I usually put four ladies up one tree and four up another. You get to know the trees after all this time, you know where to go with the ladder. There's some grand trees."

I asked the women if they could move the ladders themselves, and they laughed at the idea.

"No way! Let Maurice strain his guts, eh? . . . ." "He's not a bad guy, anyway, but he'll send you right back up there if you've missed some."

"They're about forty-five rungs, these ladders," Maurice said, "and you've got to keep them straight when you move them or you're in trouble. It can be tricky in a wind. The men over there are outsiders, and they get paid that bit extra for moving their own ladder. They get £1.30 a tray and the ladies get £1.25. And they'll pick 14 trays each on a good day. That's good money in a good year like this, but then it varies with the crop because it's the same rate whether there's lots of cherries or whether you've got to hunt for them. This is one of the best years we've had. And then of course they eat a load, though I never. I eat one cherry first thing and keep the stone in my mouth all day for luck."

I asked Maurice if the birds had been a problem.

"No, we've done well to keep them out this year. No thanks to those bangers. The birds don't take the least bit of notice of them. You've got to have a twelve-bore shotgun and you've got to keep going . . . two of us, all day, when it's called for, then you can keep them away."

I asked him if the weather had been a problem.

"You're joking! With a summer like this? Mind you, if it rains too soon now we'll have a lot of split fruit. We've got a lot to do yet, so we don't want any rain."

But there *were* problems with the cherries? All these orchards would be gone in a few years?

"Well yes, that's what they say." And he hunched his shoulders philosophically without breaking his smile.

14

The Apple of our Isle — Apples and Pears in Kent

PLATFORM 8, VICTORIA STATION. Eight a.m. A liveried footman wearing white gloves opened the door of the carriage named 'Audrey' and I stepped aboard the Orient Express. Fifteen minutes later I was served a champagne breakfast as the historic train sped through southeast London and out into the Kent countryside on its way towards Dover.

I was not embarked, however, on a trail of intrigue or espionage in the company of a recently widowed millionairess or an illicit arms salesman, with an urgent rendezvous in Istambul. I was a guest of the Apple and Pear Development Council, along with ninety-nine other journalists, from the commodity-market analyst at the *Financial Times* to the cookery writer on *Nursery World*. The 'Trail of Discovery' took us out to Teynham, and the part played by the train in publicizing the arrival of the new English season with Discovery apples was considered cheap at around £10,000. "That would only have bought us one minute of TV time," an organizer told me. "But it leaves us with a problem of how to follow the act next year." The APDC has developed quite a flair for publicity since its creation in 1980. With executives dressed in Restoration costume at the Theatre Royal, Drury Lane, they rewrote history one year with Nell Gwyn selling not oranges, but orange pippins. They regularly sponsor Christmas wassailing, when a group of children from the Brenchley Village Choir sing and present apples to the royal palaces, offices of government, or foreign embassies. And contrariwise, they present a floppy cuckoo award from time to time, when growers market immature or otherwise substandard fruit.

From the train we were taken to the Bouchers' Newlands farm, which I had visited a few weeks earlier at cherry-picking time. In the large, open yard under a hot sun, Rex Boucher called out to the cohorts on his left and then the cohorts on his right as he gave the assembled crowd of writers and broadcasters a brief introduction:

"*'Deus dat incrementum'*, or 'God gives of the increase'. . . . The sky is
our roof and we take what is sent to us, that is a salutary given of those
who work in the fresh air. . . . We have a good climate, and a good
alluvial soil which is deep, to retain moisture. . . . The amount of
minerals taken out of our soil here is minimal." He took a swipe or
two at his French rivals, "In the Rhone Valley it's solid sand and they
have to put fertilizer on their Golden Delicious by the bucketful . . .
they have to water them all the time . . . and they taste awful!" He
mentioned a recent article in the *Sunday Times* which reported on
pesticide residues in fresh fruit. "He got his facts all wrong. . . . Of
course we like to keep flies out of the kitchen and woodworm out of
the loft. . . . The leaf is our factory and we have to keep it hygienic.
We work out there for six months when the leaf is on, and we want to
stay healthy. . . . We use the very minimum of insecticides." He
smiled generously. "The men and women who work on these farms
are my friends. They're a marvellous staff . . . I say good morning to
the trees. . . . They think and feel. . . . Cox is a remarkable variety
but they're just like women, they need daily attention or they might
take off. . . ." There were some titters here from his audience. "The
housewife is the boss and we never forget it. . . . This industry is
vibrant and if anyone can show me better apples, I'll grow them." He
finished with a cheerful flourish, "The essence of fruit is enjoyment
and good health."

I set off on a tour of the orchards with a portfolio of printed material,
in a group led by Rex Boucher's son, Hugh, and including several
other top fruit growers who were available to answer questions. First,
a 9-acre orchard of Bramleys.

"These trees were planted in 1967 on the M26 root stock with a
Long Ashton Research Station selected clone. There are 93 trees per
acre and we estimate a crop this year of 10 tons per acre. For
pollination we use one-in-nine Grenadier, which have already been
marketed, and various strains of Malus — that's the crab apple — as
trees, and as grafts on the other trees. . . . You will notice that a heavy
'June drop' took place. The trees are intelligent enough to know what
quantity of fruit they can successfully carry. . . . Some apples show
sun scald due to the strong sun in early July when the skin was fast
growing and tender." The trunks of these Bramley trees looked like
those on the old standards at the bottom of my garden which I climbed
up into as a child and shook ferociously to dislodge apples way above
me. But about five feet from the ground something drastic had been
done to them by way of pruning or pollarding, so that the branches

fanned out horizontally, bending down at their tips almost to the ground under the weight of fruit, so that the tree looked like a ragged umbrella.

Then past a windbreak formed by a dense line of poplars and alders to a pear orchard, again planted in 1967.

"Ideally we should have windbreaks every 120 metres. They create microclimates within the orchard and by reducing air movement in spring they can raise the temperature for the blossom by as much as two degrees. And if you have high winds tunnelling down in August and September you can lose a quarter of your harvest." Pears flower earlier than most apples and perform markedly better under the protection of shelter and close planting.

"With the pears we have a tree population of 792 per acre. The Comice yield 5½ tons per acre, the Conference 15 tons per acre. In 1981 a major change in pruning policy was felt necessary, as the growth in the tops of the trees was becoming too dominant and the shading of the lower branches was causing loss of yield. The trees were cut very hard in the winter of '81–'82 and a lot of detailed tying down of young shoots was undertaken. Heavy pruning puts the tree under stress and gives you big buds. It's like any plant, trying to reproduce itself if it's threatened — they reproduce better under stress. We brought an 'A' shape to the trees, taking out top wood and letting light into the lower branches. The young branches were taped down for two years to fill the fruiting canopy. Then the fruiting wood was at an optimum and from then on we just cut half off the new shoots. The increased light available to the trees seems to have had the desired effect and yields have increased by up to a third from 1980 levels."

The orchard of Discovery apples, which we had especially come to see, was about 5 acres, with a yield of 7½ tons per acre. "This was planted in 1975 at a planting distance of sixteen feet by twelve feet, giving 227 trees per acre. The root stock is M26 and pollination is by four types of Malus and in-tree grafts of Cox and Idared. We get 50 hives of honey bees down from Cumbria at blossom time, but some growers worry about them carrying problems in, and rely on the wind and natural insects. Discovery is generally slow to start cropping but yields have built up in the last couple of years. The farm output is approximately 95 tons, or 7,500 boxes, of Discovery and these are all marketed fresh over a period of one month, generally from August 10th to September 10th."

These trees were more spidery affairs, hung with a quantity of fruit which seemed altogether out of proportion to the quantity of wood and leaf. They reached no higher than five feet, so that all the work could be done from the ground, and several of the side branches reached down almost to the ground, with much fruit at knee level. Having planted some two-year-old trees in my own garden a few years ago, I had tried to follow the detailed guide to pruning in my standard amateur-fruit-grower manual, meticulously counting buds on each 'leader' and side shoot. The business became extremely complicated when the new shoots refused to grow as the manual predicted, and I abandoned this method for that of my apparently successful neighbour. His advice was simply to prune to maintain a good shape to the tree, allowing plenty of space in the centre. This has seemed to work well, giving reasonably good crops, but my trees have the "good shape" of a stylized drawing from my children's picture-book version of the Johnny Appleseed story. So I asked questions now about commercial pruning. One grower in the group started philosophically by explaining that pruning used to be considered an art, but now it has become a science. He explained that in this particular orchard they probably hadn't had the time to get round to a proper summer pruning, but that with intensive pruning you grow a lot of little spurs with thick leaves, so the apples don't colour-up and ripen so well. He pointed out which shoots should be taken off now. A reporter for the horticultural press with years of experience took me on one side at this point and told me that every grower she had ever talked to had their own way of pruning, and every one of them was convinced that his method was the best. Eventually I was pointed to an account from ADAS which said:

> The centre-leader method of tree management is widely adopted by commercial fruit growers. The best way to control the vigour of a tree is to let it crop, therefore the basic aim in modern orchard systems is to encourage early cropping, but not at the expense of producing sufficient growth and fruit laterals to fill adequately the allotted space. The skill lies in keeping a balance between cropping and growth. In the past, there has been a tendency to over-prune fruit trees and it is important not to confuse the training of young trees with pruning.

From the orchards, Hugh Boucher took us to the nursery. "In 1982, due to the unbalanced age structure of the orchards on this farm it was decided to embark on an extensive replanting scheme. With modern intensive planting systems of up to 1,100 trees per acre envisaged, we sought some way of reducing the capital costs involved. A feathered maiden tree will cost about £2.20 from a nursery, whereas we can produce a similar tree here for half that price. We buy in root stock which has been grown from cuttings from a certified mother tree, and plant them here during January and February. This root stock determines the size and type of tree, and the ripening times. MM106, for instance, is very bushy; M26 is less vigorous and easier to control; M9 is a very popular dwarfing stock. Then in early August, virus-free Cox buds are put on. This is done by John and Sue Webb, who are specialists, on contract. They take a slither round the Cox bud and match the cambium layers with a cut on the root stock. They put tape round to keep the moisture in, and over six weeks the bud will callous-in. It's very fine work, but they just chat along as they're doing it. They over-winter and then take off growing the next year, with a 95 per cent success rate. In February or March when the tree is truly dormant we cut off the root tree bit above the bud, and by the following winter the trees are planted out in the orchard." The soil around these bandaged babies, with stems no thicker than my little finger, was entirely clean of weeds, and Hugh easily found a recently abandoned Cox branch, carefully stripped of buds, to show us.

At last we came to the cold storage complex, whose replacement value would be £400,000. The early apples are usually sold straight from the tree, but most of the 600 tons of Cox, 100 tons of Bramleys and smaller quantities of Spartan, Egremont Russet, and Red Delicious are stored after picking in late September and early October, to be sent to market continuously from mid-November to as late as the following April. We entered a cavernous refrigeration chamber where bulk bins are stacked by fork lift, eight high.

"The object is to keep the fruit in a living condition whereby respiration takes place and the apples breathe in oxygen and breathe out carbon dioxide. The chambers are sealed, when full, using air-tight doors, and the apples use up the oxygen until the level moves from the 19 per cent in normal air down to 2 per cent. This is then kept constant. Meanwhile the carbon dioxide is removed down to a level of 1 per cent or less. Also the temperature is quickly reduced, to 38°F for Cox, to prevent further ripening. This is called low oxygen storage under a controlled atmosphere, and it is very effective for our varieties."

The botanical family Rosaceae includes, as well as the rose, many of the important tree fruits of the temperate zones, such as the pear, plum, cherry, apricot, almond, peach and apple. The apple belongs to the genus Malus and grows wild over much of Europe and Western Asia. Three particular species of wild apples have been especially important in the ancestry of our cultivated apples: *Malus silvestris* which has sour, shiny green fruit and smooth leaves, and is native to

Britain; *Malus pumila*, which is a smaller tree with sweeter, coloured fruit, and leaves which are covered with down on their underside, which is more common in southern Europe; and *Malus baccata*, the Siberian crab, which produces bright red, cherry-sized fruit on a forty-foot-high tree, is very winter hardy, and resistant to scab. Similarly, our pears belong to the genus Pyrus and have descended from several species which produce small, hard, gritty, sour fruit in the wild over a large area of Europe and Northern Asia.

Whole woods of wild apple and pear trees have been found on the lower slopes around the Caucasian mountains, and frequent hybridization between the species has occurred naturally for many thousands of years. Carbonized apples dating from 6,500 BC have been found at archaeological sites in neighbouring Anatolia, and remains of two varieties of apples which had been sliced and dried for winter storage between 3000 and 2000 BC at a prehistoric lake dwelling in Switzerland. Impressions of apple pips in Neolithic pottery fragments found in Wiltshire suggest that apples were also a part of the prehistoric diet in Britain. But it is difficult to tell at what point cultivation began, for these could all be remains of apples gathered from the wild. Edward Hyams supposes that apples and pears were first cultivated in Anatolia or some nearby region where they grew wild, and which was also close to the ancient civilization of Sumer where farming and plantation agriculture were developed before anywhere else in Europe. But he also suggests that an independent domestication was achieved, albeit later, at the early lake dweller sites in Western Europe:

At first the people of those settlements simply gathered and ate the wild fruits; apple and pear trees sprang from the seeds which were scattered in rubbish and excrement, or spat out, in the neighbourhood of the village, and thus the first orchards were planted unintentionally. But the ascent of man since the dawn of his reasoning power has been to a great extent a tale of taking intelligent advantage of accidental discoveries: the vestiges of fruit found on Bronze-Age sites show a marked improvement, notably in size, over the fruit found on Neolithic sites, which can only mean that deliberate selection had been at work. Trees bearing the best fruit were preserved and propagated from, while the poorer ones were felled. The process was progressive because, given this measure of segregation of improved forms, seedlings would tend to include an increasing number surpassing the parent trees in desirable attributes.

When the Greeks, and then Romans, came into contact with the more advanced 'nurserymen' of Asia Minor, the earliest improved apples would be brought west, subsequently to hybridize with local varieties to give even more and better types. The selection of superior mutants and hybrid seedlings over such a long period has led to the existence of many thousands of different varieties.

It is because the genetic composition of cultivated apples is so complex that seedlings grown from pips may bear no resemblance to the apple from which they came, or to any other seedlings grown from fruit from the same tree. Some chance seedlings will prove superior, the vast majority will be inferior and perhaps useless. Apple trees are also difficult to propagate from cuttings, and the technique of budding or grafting the desired variety on to dubious seedlings was mastered at an early stage. Instructions given by Roman writers differ little from some methods still practised today.

The Romans introduced new varieties of apples and pears to Britain, and planted many orchards. During the Dark Ages it is unlikely that much cultivation of orchard fruits continued, although some trees and their seedlings would survive. The early monasteries planted new orchards with varieties again imported from the continent, and this process was greatly accelerated after the Norman conquest. In his book, *Cultivated Fruits of Britain*, which is likely to become a standard reference, F. A. Roach gives a detailed history of apple- and pear-growing through successive periods of prominence and decline, naming the particular varieties which were commonly grown at each stage. Many of these are still in existence, and are maintained in a kind of living museum at the National Fruit Trials at Brogdale, near Faversham, only a short distance from the Bouchers' farm at Teynham. The oldest of the 2,000 or so varieties of apples grown at Brogdale is Decio, believed to date from around AD 450. There are several varieties from the thirteenth century and then many from the sixteenth century onward, when Richard Harris, fruiterer to Henry VIII, imported grafts of many new varieties of fruits from France and the Low Countries, and planted large orchards around Teynham.

Kent has always predominated in producing apples and pears because of its favourable climate and soil and proximity to London. But orchards also flourished in Essex, in the southwest, and especially in Herefordshire and Worcester, serving towns of the midlands and producing vast quantities of cider and perry from varieties especially selected for the purpose. And by the eighteenth century apples were grown in private gardens in most parts of the country.

It was during the nineteenth century that many of the famous varieties grown widely today first appeared, often as chance seedlings taken up by amateur gardeners. Sometime between 1809 and 1813, for example, Mary Anne Brailsford planted a pip from an apple she had eaten in a pot on the window-ledge of the cottage in which she lived with her widowed mother in Church Street, Southwall, near Nottingham. The seedling was subsequently planted in the garden, and the property taken over by one Matthew Bramley. The tree produced a bountiful supply of large attractive apples which cooked to perfection and kept from October to May without special storage. Mr Bramley sold grafts to a local nurseryman who exhibited the fruit in London in 1876. Over 100,000 tons of Bramleys are now marketed each year, and they all derive from that single tree. The original still exists, and crops well. Amateur enthusiasts spray, prune and feed it, and although Bramleys are grown only in Britain, pomologists come to view this tree from all over the world. The current owner of the cottage has been known to ask them to bow three times before it.

In 1825 a retired brewer called Richard Cox planted a pip from a Ribston Pippin — itself considered by connoisseurs at the time as the best apple to accompany a fine claret — and raised a fruit which many experts believe has not been surpassed for its superb blend of rich, aromatic flavours. Grafts of Cox's Orange Pippin from the large garden at Colnbrook Lawn, near Slough, were sold by Berkshire nurseries and the apple quickly became popular with gardeners, although commercial growers soon experienced cultural difficulties such as susceptibility to scab, and yields were not always high. The original tree is believed to have blown down in 1911, but its descendants now yield well over 100,000 tons in the United Kingdom in a good year.

Worcester Pearmain was similarly a seedling of the old west country variety, Devonshire Quarrenden, raised by an amateur near Worcester in the 1860s. It soon became famed for its brilliant colour and its heavy, regular yields.

Early this century a new range of varieties was created through controlled hybridization by professional plant breeders, in the hope of combining the better qualities of two existing varieties. Young trees grown in pots would be isolated in insect-proof greenhouses when in flower and, with stamens removed to avoid self-pollination, the trees would be cross-pollinated by hand. When the fruit had grown and matured, seeds from it would be sown and the resulting progeny awaited with hope and perseverance — for each offspring of promise is accompanied by thousands of useless ones. The Laxton Brothers of

Bedford crossed Worcester Pearmain with James Grieve to create, eventually, Lord Lambourne; Cox's Orange with Wyken Pippen to give Laxton Superb, with a Cox-like flavour but capable of storing easily into the new year; and Cox's Orange with Wealthy to give a very juicy apple with a Cox-like flavour, Laxton's Fortune. At the East Malling Research Institute H. M. Tyderman created Tyderman's Early Worcester and Tyderman's Late Orange by controlled hybridization.

Various new varieties were also imported from abroad. Golden Delicious was a chance seedling found in West Virginia in 1890, and this was crossed in the 1930s with a Japanese apple called Mutsu to create the excellent crisp, juicy and sweet Crispin; Ingrid Marie came from a chance seedling of Cox in Denmark; Katie from a Swedish cross-breeding; Kidd's Orange Red and Gala from New Zealand.

These varieties are all included in a list of 50 described in a leaflet, *Take a Fresh Look at Apples*, published by the English Tourist Board, together with the addresses of farms where they are still grown and available for sale. On this list, too, are the names of many apples which I remember making a fleeting appearance in my father's warehouse when I was younger — such as Beauty of Bath, George Cave, Miller's Seedling and Ellison's Orange — which were superb at their peak.

Much of the modern breeding, which continues apace at East Malling, is concerned with creating varieties which crop heavily and regularly, and which will store well. There is an increasing commercial concentration on Cox and Bramley, and according to the APDC these will remain the dominant varieties in the UK well into the next century. But there are still openings for other varieties. Discovery, which began life as a chance seedling in an orchard of Worcester Pearmains near Colchester in the late 1950s, has become the third largest apple variety in terms of area, with its special status as the first apple of the season. There is an active search for good varieties to bridge the gap between Discovery and Cox, presently filled mainly by Worcester Pearmain and Katie. Spartan currently adds the choice of a red apple to the mid-season Cox, and there is a desire to extend this availability of highly coloured fruit. There is an ongoing search to find improved clones of Cox giving us Queen Cox, Cox Emla and Flanders Cox so far, and there is also a need for commercially-attractive pollinator trees. In a Cox orchard about one tree in six is planted to pollinate the Coxes, and many of the varieties used for this in the past have borne fruit which is unattractive itself. So there is a future here perhaps, albeit only second to Cox, for a newcomer like Elstar (Golden Delicious. x Ingrid Marie), Kent (Cox x Jonathan), or

Jonagold (Jonathan x Golden Delicious). Once a newly created variety has left Malling Research Station it undergoes lengthy commercial trials at Brogdale. And then the APDC conducts a series of detailed and widespread tests for consumer response. Several new varieties such as Jupiter, Suntan and Greensleeves have been hailed prematurely by the press, only to fall foul at a later stage of development.

On a parallel course with the development of our apples is that of cultivated pears. During the revival of orchard cultivation in the thirteenth century the Wardon pear was raised by Cistercian monks at Wardon in Bedfordshire. This was a large baking pear — the inspiration for Wardon pies — whose reputation grew to such a degree that 'wardons' were subsequently considered a distinct type of fruit. By the early eighteenth century a wide range of varieties of pear were grown, with over 600 listed by the Royal Horticultural Society, in 'summer', 'autumn', 'winter' and 'cooking' categories. Pears were often grafted on to hawthorn, but now special dwarfing quinces are used as the root-stock. Williams' Bon Chretien still sold as 'Williams' today was bred by a schoolmaster at Aldermaston in 1770. This became the Bartlett pear when grown by Enoch Bartlett in Massachusetts and is still one of the most favoured pears for canning. Doyenne du Comice, much valued for its flavour, was raised from seed at Augers in 1849, following Beurre Hardy at Boulogne in 1820. Then in 1895 Thomas Rivers entered a new pear — raised from the pip of a cooking pear called Leon le Clare de Laval — at an International Pear Conference. The competition judges liked the pear so much that they asked it to be named 'Conference' in honour of the occasion. Most considered opinion now accepts that Conference is not a pear of outstanding quality. It is well suited to our climate, however, and is a regular, heavy cropper. Picked unripe, it will also keep for many months in cold store, replacing the old varieties which were naturally long-lasting. Conference now overwhelmingly dominates the pear harvest. And yet the acreage of our pear orchards has been greatly reduced over recent years. Pears are now strictly an 'impulse buy' and, at about 3 lb. per person per year, Britain has the lowest rate of pear consumption in Europe. There is only one real new contender for a place on future markets, which is a cross between Conference and Comice called Concord. "With roughly double the yield of Conference," says one of its developers, "there are some minuses . . . such as its lack of flavour."

Beyond the work with new varieties and root stocks — which now enable Bramleys to be available all year round — there is an ongoing change in techniques of culture. It was knowledge of the

nutritional requirements of apple trees and increased disease control
which led to much increased planting of Cox in the 1930s, and
modern calcium treatments are gradually eliminating bitter pit in
Bramley. This shows as faint underskin bruising on the outside, with
small brown spots throughout the outer flesh of the apple, and it is
hoped that its total demise will help to restore Bramley as a first-class
culinary resource and "British staple food". Similarly, it has been
common practice for many years to use herbicides under the trees,
with closely mown grass in the centre of the alleyways where the
machinery runs. But only quite recently it has been discovered that
water stress not only affects the current crop, but the cropping
potential for subsequent years. So the removal of competition by
weeds is now often supplemented by carefully controlled irrigation,
and sometimes fruit thinning. The latter can be achieved using
chemicals, as can the improved initiation of fruit buds. Hormone
treatments can be used to get a crop after frost damage to the
blossoms, to speed up the colouring and ripening of the fruit, and to
slow down unwanted growth of wood. And sophisticated spray
programmes can so minimize the damage caused by pests and disease
that the modern consumer can be supplied with uniform, blemish-
free fruit which he or she evidently values.

One of the hosts on the Orient Express 'Discovery Trail' was Mrs
Teresa Wickham, who serves with seven other growers on the
council of the APDC. I went to visit her later at her own farm at
Brenchley, six miles south-east of Tonbridge on the edge of the Kent
Weald, and went for a delightful walk on her own 'Gate House Farm
Trail'. Incorporating an already existing public footpath, and in
collaboration with Kent County Council, this trail takes in much of
the 80 acres of orchard and soft fruits. There is a written guide which
includes considerable information about modern horticultural practice while also reflecting the Wickhams' keen interest in conservation. This is distributed to schools and libraries and generates a great
deal of interest, with many parties of schoolchildren visiting the farm
between April and October. This accords with another of Mrs
Wickham's interests as a founder member and then chairwoman of
the Women's Farming Union, one of whose specific aims is to
provide a link between producer and consumer for their mutual
benefit.

Gate House Farm is on the clay soil known as the Tunbridge Wells
geological series. This is a part of the Weald which consisted
historically of traditional woodland growing on the broad clays

deposited here on top of the underlying chalk. From a northerly point on the trail there is a good view across the Weald, with much of it still standing out as deciduous woodland. Forty-eight per cent of our national top fruit crop is grown in Kent, and one historic factor leading to this, in addition to the fertile soils and favourable climate, is the fact that the open field system of strip cultivation was never widespread here. Nor, after the dissolution of the monasteries, was Kent ever dominated by large landowners. Farmers often owned their own land, and were thus free to plant orchards as soon as this appeared profitable. Initially Gate House Farm was a mix of fruit, livestock and hops — and the old oast house where the hops were dried is now the farmhouse. Sheep grazed under the big old Bramleys up until mid-century, and pigs and bullocks were also raised on a small scale.

"My husband Robin was born next door, where he worked for his father until we bought this farm from him and went into partnership on our own, four years ago. So having been a manager for many years he's now a farmer in his own right. We have another unit up the road at Matfield giving us 120 acres altogether, with about 60 acres of apples. My role is more on the business and marketing side, having a food industry background. My husband adores growing. But nowadays the farmer must be a good businessman. You've got to have good financial sense as well as being a grower. You find more women are becoming financial partners on farms. A lot of our members in the WFU have married into farming. They don't come from a farming background themselves, but they want to take an active role. And as farmers cut back on labour it's no longer enough for her to be quietly answering the phone and preparing meals. It's mostly a generational change. My mother-in-law was more highly trained than my father-in-law and yet she was never allowed to be called to the farm. But now at my generation the women are much more involved. Mostly it's aligned to the marketing side, although there is a higher proportion of women at agricultural college than there's ever been before. Of course that's still small . . . and a lot of the girls go out to jobs in environmental work.

"Our major variety is Bramley, then Cox, Crispin, Spartan, Egremont Russet, and Golden Delicious and Discovery which are planted mainly as pollinators for Cox. Then we have just small quantities of Howgate Wonder, Lord Derby, Ingrid Marie and a few others. But we shall get down to just four eventually. Quite a few people are trying Katie, Kent and Jonagold. But our replanting programme so far is up to Bramley. A new orchard lasts fifteen to twenty years and we have a constant replanting programme. We have some more of our older types to get out yet and then we shall have Bramley, Cox, Discovery and a few Worcesters. The biggest change on our market has been the growth of the supermarkets, which has had a dramatic effect. These four varieties are the ones which sell.

"One farm markets through HGF (Home Grown Fruits), which is in the Kingdom scheme, and the other markets through SGT (Society of General Traders) which is not. We work on the Prestel, with our own page number. We're picking Grenadier today, for instance, and we'll be on Discovery next week. We pack these straight into HGF boxes and they will go straight to the market. They might say, 'We'll take two pallets and they're going to Cardiff market', so we have them ready by a certain time, they arrange the transport, and that's that. . . . We will put it through the Prestel, and say what we're loading and what we've got coming up, and their marketing commission comes up there, and what they've got in stock and what's shifting well and so on. Then there are the supermarket direct orders and HGF or SGT will ring us and we can bring apples out of store and pack them as required. You can be very sophisticated, with water flotation schemes and computerized control, but you can also grade by hand, like we do, quite successfully. You've got to have quality control and discipline. A lot of people pack for others and make a business of it, like my father-in-law, so he can keep his women going all winter. Because we have soft fruit — which we can grow in the frost-pockets not suitable for apples — we have a regular core of workers. So they can either be cutting out the old raspberry canes in winter or we can switch them into packing for a special order.

"Our touch with the consumer as an ordinary grower is very little. But in my other hat with the WFU it's very different, because we run an ongoing surveillance scheme. We have these postcards that go out to women all over the country and they send back weekly reports on the price, condition and availability of produce in the shops. It gives us absolutely vital feedback. We've written reports on all sorts of commodities but we started out actually as an apple action group in 1979. We had a mad apple campaign and fought the French on the beaches and all that sort of thing, and got the industry together. That's what helped to get the Kingdom marketing scheme going, which has been one of the pioneer schemes which Food From Britain has adopted for its Quality Food Mark. Our aims are to stimulate demand for high-quality British produce, to encourage better marketing practice, and to provide links between producer and consumer. We're quite independent of the NFU (National Farmers Union) — even the female branch of it! We get sponsorship and some industry funds to help with administration but no one gets paid. It was all run from this office originally. It's been a tremendous experience in seeing what goes on. . . .

"We think farming is going through the biggest revolution it's ever gone through . . . it's changing. . . . It's no good producing the stuff the customer doesn't want to buy.

"People really want flavour, and our climate enables us to grow the best flavoured apples. Cox has always been more expensive than Golden Delicious, whether it's a French Golden or an English Golden, simply because it's a more expensive apple to grow. But now Cox has established its own price patterns, independent from French Golden Delicious. So we're the upmarket apple, and that's where I think we should concentrate our effort. We can do Spartan as a slightly cheaper apple but we haven't actually found an equivalent in this country to the French Golden Delicious. We've come closest perhaps with Crispin. It's not so uniform looking but it yields well and stores well and flavourwise it's lovely. And the public appreciate that. Hence the farmshop syndrome. But in the High Street it just doesn't take. A lot of the old varieties are marvellous but the ordinary customer just won't buy them. And the other thing is, you've got to cut them fresh. A lot of growers offer Pick-Your-Own and they can indulge in older and less familiar varieties because they get a lot of enthusiasts coming out. But it's different in the supermarkets. People over thirty-five will recognize a Russet, but growers have become disillusioned with Crispin because the marketing side just isn't there. People see a yellow apple — and a properly ripe Golden Delicious from Britain or the Loire Valley can be a very nice apple — and they think it's over-ripe. We've got to grow something which our customers can immediately recognize and want to buy. Like Cox. So that our variety is our brand. It takes three years, if you're doing intensive planting, before you get any return, so you've got to have a lot of faith to make the investment. That's why people are going for the streamlined varieties.

"The trouble is that although the APDC gets £29 per hectare from the 1,300 top fruit growers in this country, this only gives us half a million pounds to spend on promotion — including the administrative side — and it just isn't enough to combat the French. One of the things the WFU has done is given evidence about the funding of the TV Le Crunch campaign and the tremendous input of the French government into that. There are natural advantages you can't do anything about. Some people might ask, for instance, if they can grow tomatoes outside in natural light in parts of the EEC, why are we spending £10,000 an acre to grow them here in heated greenhouses? Well Golden Delicious has some of those natural advantages. But so does our Cox. Our apples are better flavoured precisely because of the climate — the amount of rain and the slow growing season. And the consumer should never be deprived of that. What has to change is the level of support given, because no longer will anyone tolerate surpluses that are paid for — by the farmer as well as the consumer — that just become mountains and go into stock. In Poland they've just set up an apple-processing plant which the EEC paid for — for Poland to juice and concentrate our surplus apples!

"It's absolutely fine if natural advantages are taken into it freely and fairly, but not when you get extra government help to subsidize transport, or advertising, or gas to heat greenhouses. In France, agriculture is their green oil and it's because agriculture has this different importance in different countries that we really have to revise the Common Agricultural Policy. The support price for cereal is the craziest thing, so we've got people growing the stuff on land that's totally unsuitable. It's the politicians, in the middle of the day, who have made a muck of it.

"Because the industry is the way it is at the moment, it is dependent on a certain level of support to ensure that the country is fed with basic priorities. But the government is not giving any clear lead to farmers. Instead of coming out clearly and saying, for instance, 'We know the customer wants lean meat so we'll see that you get a premium for lean meat' . . . at the moment it is actually to the producer's benefit to produce a fat lamb, because the support is by weight. What the customer wants in a supermarket is a lamb that is not so fatty. So this discrepancy between what the customer clearly wants and what the farmer is being paid to produce is one of the major things that has got to be sorted out. Certainly, with apples, we can compete as long as the competition is fair. Because we are producing quality apples which people want."

At this point I mentioned a newspaper report on a confidential survey carried out by the "independent and well-respected" Association of Public Analysts, which Rex Boucher had mentioned briefly on the 'Trail of Discovery'. The article claimed that one-third of all fresh fruit and vegetables on sale in Britain is contaminated by chemical residues and that according to a Harley Street doctor who specializes in allergies, there is now irrefutable evidence to link the ingestion of such chemicals into the body over a period of time with an exhaustion of the immune system. And yet the chief executive of the APDC said on the 'Trail of Discovery': "In the last decade or two we have been warned off all sorts of things as one zany academic or another postulates dietetic theory. Never top fruit."

As well as receiving regular herbicide sprays, British 'top', or orchard fruit typically receives pesticide sprays against scab, mildew, and aphids, caterpillar and codlin moth. There may be chemical treatments at blossom time and towards harvest time to further colour and ripening, and to aid storage. And there is also a developing technology to extend shelf life out of storage involving dipping fruit in chemical preservatives or irradiating it with gamma rays, both of which have been used to some extent in other countries.

"The gamma rays business . . . it seems absolutely horrific to me . . . I really know nothing about it. A lot of the Americans wax their fruit, but we would be very loath to do that in this country. Our great advantage here is our closeness to our markets — if you've got your marketing sorted out you can get your stuff in fresh and that's what the supermarkets want. But this concern is building up . . . we've just really got going with this . . . because people *are* concerned and *are* worried about what they are eating, and chemical residues are going to become a much higher priority. There was that example of the M.P.s' gin and tonic, for example." She was refering to a University of London study showing that the action of the alcohol on the wax-treated skins of lemons led to a fungicide level in G & Ts containing lemons of twenty times the amount normally considered safe. The Ministry of Agriculture refused to comment, saying that the information on pesticide levels was governed by the Official Secrets Act. Teresa Wickham's comment was simply: "There's a general awareness of the problem.

"HGF reckons growers are responsible for spraying, and they have cut-off dates for when you can do it. Our apples are tested, and there are fairly strict rules. We've got four big files on marketing directions and rules. Some people, for instance, spray Discovery with a thing called 'Ethrol', though we won't use it ourselves. It gets your apples ripe a little earlier, with a little more colour. But the fruit starts to mature as soon as you spray it on, so it has a shorter shelf life. Now HGF will say: 'Yes, you can use it for your first pick over, but not after that.' So you get a certain guideline. If it's a particularly wet winter and there's something around like scab or mildew, then the HGF area adviser will be aware of it and offer advice, and take samples away to East Malling to have them analysed if we're worried. There are several groups of advisers, including ADAS.

"Most of the things we put on are insecticides and I think they are safer now that DDT has been banned, though 2-4-5-T is a question mark. And sprays are so expensive that nobody wants to put on more than they have to. I think we've got it pretty well down. But I think it's highly unlikely that we can ever grow the kind of apple the consumer has come to expect, totally by organic methods. It's not only the fruit, it's the actual tree you've got to look after. We can get things like fireblight that comes off the hawthorn, and if it gets hold, the tree dies. It's dead. There are quite a lot of things you can get, and you have to be careful.

"I think this is an issue which is going to be important. But if growers want to cut back on sprays one of their biggest worries at the moment is lack of research for things like this. The government have cut back terribly on research, and we need independent bodies like East Malling. If the cutbacks continue . . . the only guidelines we get in some cases are from the chemical companies themselves, who I'm sure are right and correct and careful, but you need impartial views."

The Gate House Farm trail incorporates all sorts of features alongside the beautiful orchards, the fertile fields of Himalaya blackberries, strawberries, raspberries, blackcurrants, and small young vineyards. There are wide hedgerows, splendid specimen oaks and numerous 'wild' areas attractive to a wide variety of insects and birds, with at least a dozen different tree species. One larger area of woodland is out of bounds to the public as a refuge for wildlife, though some coppicing is carried out when appropriate. This traditional form of management of the wood increases the amount of light and encourages a rich ground flora — including Bugle, Yellow Archangel, Golden Saxifrage, Garlic Mustard and Wood Anemone.

"We are having to grub out the hawthorn," Teresa Wickham told me, "because of the risk of fireblight. Otherwise the birds are the main problem. We have to hire a 'cherry minder' to scare them off when the fruit start to colour, or we wouldn't be left with a cherry. They also eat the fruit buds on the apple trees. We know that the first row of the orchard next to the wood will be stripped and we just accept that as a given fact. Rabbits can be a nuisance, too. We have to wire round every grape vine, for instance, to protect them. We used to have a ferreter out, but now my son is in charge of it. I've got two young teenagers and one of them is supposed to be a great shooter. But he came back recently and said 'Mum, I saw all those babies out, and I didn't have the heart.'"

I finished the trail beneath a solitary and heavily laden Bramley which could have served as a signature for all the orchards in Kent. Fleeting reflections, from Isaac Newton and William Tell right back to the Garden of Eden, were chased from my mind by the solid words of Coleridge: "A man cannot have a pure mind who refuses apple dumplings."

15

Queer Gear and Hodden Fodder — A Miscellany of Crops in the Vale of Evesham

I ENTERED THE valley of the River Avon, better known as the Vale of Evesham, from the west, along the same route as the first Anglo-Saxon settlers of the Midland Plain. On the low slopes of solitary Bredon Hill, at the mouth of the Vale, I first noticed a market-garden smallholding, and stopped for a word. Three elderly ladies sat in silence on upturned boxes, bunching spring onions in sixes with rubber bands, and then in dozens with twine. In the shadow of a small stone shed an adolescent boy was scratching with a hoe among a patch of thistle-like plants which I realized with surprise were globe artichokes. And then a wiry, vigorous man with a bent back came striding towards me, stepping over fifty-foot-long rows of winter cabbages, leeks, and kohlrabi. But we had few words, for this was a family of Italian immigrants, and the old man, at least, was not very expansive in English. He gesticulated with his arms between sky and soil, suggesting that the weather was not to his liking, and then resorted to smoothing out his bushy moustache with his agitated fingers.

This moustache, at least, seemed to be thriving under the prevailing climatic conditions, and reminded me of P. G. Wodehouse's character, whose whiskers "are of a Victorian bushiness and give the impression of having been grown under glass". The Victorians grew a host of exotic crops under glass, exporting their hothouse grapes to France and the United States, and as I drove on up the Vale an imbroglio of images from the horticultural history of the land around me developed in my mind, sparked by this first encounter.

The Romans drove two straight roads across the Midland Plain, and before them prehistoric peoples built a few hill forts on the higher ground of the limestone escarpment to the southeast of the Vale of Evesham. But the heavy clays of the Vale and surrounding plain were thickly forested with oak and ash and were virtually uninhabited until as late as the sixth century. The Anglo-Saxon colonization was a long, slow process and the great mass of forest was not effectively cleared until the end of the thirteenth century. In the eighth century, however, a Benedictine monastery was founded at Evesham, and horticulture, encouraged by the order, flourished on the rich soils. The Domesday Book records that more than half of the county of Worcestershire was in the hands of the Church, limiting the rise of a local aristocracy. Onions, leeks, beans, peas, cabbages, kale and parsnips were grown in farm as well as monastic gardens, and fruit orchards flourished, as they still do, in the west of the county. In addition to the fertility of the soil, the Severn Valley served as a channel for warm air from the south. There were 38 vineyards in the vicinity of Evesham in 1086, although this industry declined over the following centuries as a long-term cooling of the climate occurred nationally.

Evesham Abbey was destroyed at the Dissolution, but soon afterwards market-gardening "received an impetus from an Italian gentleman, one Francis Bernardi, agent from Genoa in this country. Objecting to some measures adopted by the State of Genoa, he resigned his post and retired to Evesham." He invested the then phenomenal sum of £30,000, far outreaching the latterday investment of Italian growers. These arrived in the Vale in some numbers in the 1950s (along with a few Pakistanis), brought in by larger growers on special land-workers permits when labour was hard to find, and buying a bit of land for themselves if and when they could.

By 1768 Arthur Young found in Evesham that "The employment of poor women and children is chiefly with the gardeners of whom—as at Sandy in Bedfordshire — there are great numbers. The produce is carried to Birmingham, Worcester, Tewkesbury, Gloucester, Warwick, Coventry, Stow etc. . . . and asparagus to Bath and Bristol." Another correspondent at that time described Evesham as 'The Eden of England in respect to gardening", although local seedsman James Bigg in his 1791 catalogue stressed the international pedigree of his wares, which included: "The Ethiopian Branching Broccoli; Yellow-seeded American Cos-Lettuce; Large leaved African Endive; New Malta Cauliflower; and The Levant Cucumber." Most gardeners, in fact, saved their own seed from selected plants to develop their own improved strains. Among the less common crops grown were several acres of "poppyheads for sale to druggists". And during the First World War there was an urgent planting, over dozens of acres, of medicinal herbs such as belladonna and blessed thistle.

In the middle of the last century several new developments occurred leading to a great expansion of market gardening in the Vale: direct rail connections with the rapidly expanding industrial cities opened up new markets, and facilitated the sending of highly perishable produce as well as bulky items such as cabbage. At the same time, large amounts of stable manure and soot were brought back from Birmingham, together with other fertilizers such as dried blood, bonemeal, and hoof and horn from the city slaughterhouses; fishmeal; and leather waste and shoddy from various manufacturing centres. A more secure system of land tenure also gradually developed whereby departing tenants recommended to their landlord a successor who agreed to purchase all the improvements such as fruit trees and other crops already planted, even giving credit for the potential value resulting from efficient working of the soil and the unexhausted manures it contained. Known as 'the Evesham custom' this enhanced the capital value of the land and encouraged the further development of intensive cultivation. Market gardening spread rapidly to field scale, and by 1906 covered some 20,000 acres. 'Warm houses' were used to raise many young plants, such as the tomatoes which were grown out-of-doors on about 250 acres, and glasshouses proper were built on many holdings.

When I was a boy we received huge quantities of produce from the Vale of Evesham on the Lancashire wholesale markets. It travelled by road overnight three or four times per week virtually year-round and included an enormous range of fruits and vegetables. But since the 1960s there has been a decline, and with good support prices for corn, market-garden land has increasingly become cereal farm land. The demand for many of the crops in which Evesham growers excelled, such as garden peas and green beans, has collapsed, and there is no local processing industry of any consequence which could compensate for this. And although the oldest marketing co-operative in Britain is Littleton and Badsey Growers Ltd, formed near Evesham in 1908, local growers have not generally combined to create a marketing force capable of holding their own with the rise of the supermarket sector. Whereas 65 per cent of vegetables grown in Lincolnshire, for example, are now sold through co-operatives, the typical Evesham grower has developed a reputation in the trade as something of a Victorian. "He wants to go his own way and be his own boss. He likes to spread his risk over a range of crops, instead of specializing, and over the year, so he's always got a bit of money coming in. In other words, he's old-fashioned."

I found both confirmation and denial of this reputation at Amberley Farm, just outside the town of Evesham to the south.

★

Peter Hoddinott speaks with a carefully measured, deep-toned voice which gives an impression of maturity and cautious confidence. Anyone would think twice before calling him an impetuous hothead. And yet Peter is only twenty-four, still studying for his B.Sc. in Agriculture and Horticulture. He is brimming with new ideas, gleaned from the three-year sandwich course which gave him his Higher National Diploma in Commercial Horticulture; and he is eager to put them into practice, working in parallel with his more conservative father. He showed me round his large polythene tunnels full of Capsicum and chillis, and glasshouse of aubergines:

"I became interested in these alternative crops when I was working at the Luddington Experimental Station near Stratford, on the sandwich course. We found one variety of Capsicum — Serina — was very good under tunnel conditions, whereas Shamrock was the best under glass. They're both US varieties, and they've come out tops over three-year trials. Of course, they might be superseded in the near future, we don't know. Father is quite doubtful to change but I've come along with new ideas for the business like the aubergines, and we've had chillis for three years now. The first year we purchased spare plants from the Experimental Station and just tried them out and Dad has allowed me to implement these ideas. But now we've got a good market for them. The thing with chilli plants is they don't produce any more chillis until you take off the flush of fruit which is ready. They just sit there and the chillis will go to their fully mature state, which is red. But this year we've kept up with production and our wholesalers will ring up and take everything we've got. We send them to Glasgow, Newcastle, Birmingham and Bristol. Okay, we only grow a very limited quantity — just one bay and a half in a greenhouse — but it's working out. You see, tomatoes have become very high input, using artificial media like rockwool or NFT for maximum yields, and with old-fashioned glass there's no real control over the environment. With this low-input glass I think that alternative crops like chillis are the answer.

"There is some new glass around, over at Badsey . . . modern Venlos. . . . We do have some advantages here. We may not have the light levels that they have on the south coast but we're certainly no worse than the Dutch, and look what they achieve! We're also very well placed for access to the major markets, and there's a really good network of transport from Evesham. I used to go down to the station with my Dad when I was little, but now it's all by road. It's such a good network that import-merchants have set up here, to distribute to the rest of the country. There'll be two or three continental lorries a day going into the industrial estate. Foreign flowers will land at Heathrow, for instance, and be brought to Evesham for distribution.

"Then there is the soil, although with tomatoes there is very little grown in soil these days. The soil on this farm is quite a heavy loam, but some land around the farm is a little bit lighter which is why we have an extensive Brassica and cereal rotation, which suits it. If you've got a really light soil like they have in the eastern counties, you have to irrigate and irrigate and all it's doing is leaching away your nutrients. They haven't got the bulky organic material — the farmyard manures — to keep the buffering of the soil, the water-holding capacity and the nutrient-holding capacity. We use farmyard manure every other year in the tunnels and glasshouses. We get it from a farmer down the road. We have to do the work of clearing out the cattle sheds, and there is a cash transaction too. Father has done it for years and years. It does a world of good.

"I think some of the people who have got jobs in horticulture haven't got the basic growing skills.

"Father is in charge of all the outside crops, on about 60 acres. He's got 5 or 6 acres of sprouts and he grows lots of other Brassica. He's got leeks on the market for about six months. They're busy picking ground beans at the moment, and the pole beans will be ready in another week. There's always something. You can adjust the drilling machine to suit the crop, but when it comes to harvesting and washing and grading you've got to be more specialized. You can specialize in just a few crops or grow a range that require the same machinery. But often old Evesham growers are difficult to change. The greenhouse work fits in with college, and I can do my own thing.

"I've been offered various jobs, but I could only get so far . . . the ADAS people told me to get a degree first and go on from there. I would like to do research, but if I got bored in a chemical company I'd like to come back here. I don't think I could take a job with a chemical company or anyone else without having my own plot of land to just potter around on at the weekends and so on. I'd have to have a greenhouse to grow a few tomatoes and peppers in, even if it didn't make any money. Of course, there can be hard times on the land, and Dad works all the hours God sends. If I can pick up a really good job here or abroad — there's a lot of interest in horticulture in the Arab countries, for example — I could perhaps make a decent bit of money to set up properly with some good capital. But I mustn't be too optimistic, or I'll probably end up on the dole queue for a year!"

Out in the fields surrounding Evesham and the nearby villages of

Offenham, Littleton and Badsey I saw plenty of the old-fashioned, bread-and-butter crops. Cabbage remains one of the most important. I saw round Primo and pointed Hispi types, ready for cutting; winter drumheads, red cabbage, crinkly savoy and blue-tinged January King just beginning to form into heads; and amorphous rows of green leaf which would become the spring cabbages of next year. All of these are grown on a larger scale, as farm crops, in Lincolnshire, Lancashire and Kent, as are cauliflowers and brussels sprouts. But the Brassica remains an important stand-by for Evesham market gardeners, and they capitalize on their advantages whenever possible. They invariably send the earliest sprouts to market — in August — for example, and their spring greens are more dependable in hard winters.

There is a relatively new Brassica crop in the Vale, introduced from Italy although not, as far as I could discover, by any of the growers in the Vale who came from there themselves. This is calabrese, which belongs to the sprouting broccoli branch of the cabbage family. Broccoli takes its name from the Italian *brocco*, meaning shoot, and with the exception of the Cornish growers who use this name for their winter cauliflower, it commonly refers in this country to purple sprouting broccoli. This crop is usually sown in seed beds in the spring, planted out during the summer, and harvested the following spring, sold with a fair amount of succulent leaf around the tiny sprouting flowers. Calabrese is named after the Calabria region of southern Italy, where it has been grown since the Middle Ages. It is a quick-growing summer crop, not frost-hardy, and is usually drilled straight into the field. It forms a larger central green head as well as numerous smaller side shoots or 'spears'. The Americans call this 'broccoli' and grow it in phenomenal quantities. It is a highly nutritious vegetable and has largely replaced cauliflower in the USA. Some growers believe it has a similarly bright future here, and in Kent especially they have developed its cultivation on a field scale. Successive plantings using varieties such as Cruiser, Corvet and Skiff aim to give a continuous harvest through the summer and autumn but unpredictable weather in high summer, even with well-regulated irrigation, can lead to gluts followed by shortages. Once mature the heads and spears must be cut promptly, for in a matter of hours in a warm, humid climate the tight green florets can open into a bright yellow display of unsaleable petals.

I also saw very small quantities of oriental Brassica. Most gardeners' seed catalogues list several of these types, with names like Pe-Tsai and Pak-choi. The Dutch and Israelis pioneered commercial sales of these crops in Britain, using the name Chinese cabbage and then Chinese leaves. They tend to grow very quickly, some varieties in as little as thirty days from planting, and, if planted before June, run to seed instead of producing tight heads. No doubt there is scope here for the attention of the plant breeders and experimental research stations.

A more traditional green enjoying something of a revival in the Vale is spinach, although this too came from points east originally. Its wild ancestor grows from the eastern shores of the Mediterranean through to central Asia. The ancient civilizations appear not to have eaten it, for it was across North Africa and through Moorish Spain that it first reached Europe, with the Spanish name *espinaca*, the French *épinard* and the British *espinach*, when it was first mentioned by herbalists in the sixteenth century. Round-seeded spinach with very tender, succulent leaves is grown mainly as a spring crop in the vale, although on a worker's allotment at one nursery I did see some perpetual spinach, or spinach beet, which can be cut and cut again throughout the year.

The ancestral beet is a small-rooted plant which grows naturally wild on many seashores around Europe, including Britain. The Romans, and no doubt other early peoples, ate the tasty leaves of this plant, but it was not until the fourteenth century that farmers in Germany took it into cultivation and began to select the better plants, to develop over many years various special characteristics. One avenue of this development gave us spinach beet and Swiss chard; another led to the large, coarse, yellow mangold, or mangel-wurzel, once widely grown as animal fodder; another to the white-rooted sugar beet which now provides about half of all the sugar we consume in Britain; and yet another to the distinctly flavoured, deep-red beetroot. Introduced to Britain in the sixteenth century, beetroot was cultivated at first "only in noblemen's gardens". This vegetable was destined to become perhaps the most infamous hallmark of the school canteen, but is undergoing a revival of popularity and now finds its way into a broad range of distinctive dishes. Much of our national crop is grown on the light, peaty soils of Lancashire and Cambridgeshire; harvested, because the roots 'bleed' easily if bruised, by special top-lifter machines which pull it up carefully by the leaves; and then stored for winter use either in barns with forced draught ventilation or in simple clamps outside, with straw and polythene protecting the sunken mounds of roots from frost. In Evesham most beetroot is grown for the summer and autumn markets, and some early crops are harvested in bunches with foliage.

Other roots definitely in the alternative class are salsify, once called the oyster vegetable because of its flavour, and the related black-skinned scorzonera. These long, thin roots also bleed very easily, and are peeled after cooking. Native to southern Europe and introduced to

Britain during the Tudor period, they were sometimes preferred to carrots and parsnips by the writers of the early herbals, but are now grown in very small quantities. I am not aware of any trend towards increased production.

Turnips are much older in cultivation, being grown by the Sumerians on some of the very earliest of farms. They were brought to Britain by the Romans and seem to have been an always-present though unsung part of our diet ever since. In the eighteenth century a yellow-coloured and much larger hybrid called rutabaga was introduced from Sweden, commonly called the Swedish turnip or simply Swede. These are grown widely in Northumberland, Lancashire, Yorkshire, Norfolk and Devon, where they also serve as food for sheep. Swedes grown on the red sandstone soils of Devon are considered especially fine-flavoured, and often command a premium.

With Evesham's baby white turnips and kohlrabi, or turnip-rooted cabbage, we are back in the Brassicas, all of which are prone to develop club-root in overworked land. When I came to a holding with not a Brassica in sight, the grower explained: "It's all got finger and toe — that's what we call clubroot here — and it's a hell of a job to get rid of it. They say build up your pH with lime, but it's the dickens to get rid of. So leeks are important for winter work."

Leeks, like turnips, were cultivated more than three thousand years ago in the cities of Sumer. The Roman Emperor Nero ate leeks — botanical name, *Allium porrum* — under the impression that they cleared his voice, and was nicknamed Porrophagus as a result. When King Cadwallader gathered his forces on the eve of battle with the Saxons in the year 640 he urgently needed some distinguishing mark to avoid mistaken mayhem among his own various bands of men. Leeks were grown so commonly then that the phrase 'leac tun' was used by the Saxons to describe the cottage vegetable plot. The Cymry duly raided such a plot and, each sporting a leek in their hats, went on to rout their enemy.

Some Evesham growers sow seed under glass in February, or outside in March, and then in June plant out the young leeks nine inches apart with about three inches of the elongated bulb underground, to exclude light and turn it white. But most now precision-drill the seed where the leeks are going to grow, using self-bleaching varieties with long shanks. This avoids getting soil into the leeks and better satisfies the EEC regulations which favour very white stems. The frost-hardy late winter leeks, such as the Empire varieties and Winter Reusen, tend to be greener, with a better flavour, whereas some of the early-maturing leeks I saw had fully twelve inches of white stem, with only two or three inches of 'flag' on top . . . from Celt to Mohican. . . .

Another crop ancient in cultivation, though now rarely enjoyed fresh in Britain, is the broad bean, whose season I had missed in the Vale. These beans appear to have originated in the Mediterranean basin, and wild plants still grow in the Algerian highlands. The Egyptians, Greeks and Romans grew broad beans but considered them a food suitable only for peasants. One explanation for this is that some people in these regions suffer a disease called favism whose symptoms of high temperature and jaundice are triggered by eating broad beans, or inhaling the plant's pollen. There is evidence that an inherited biochemical abnormality of the blood found in some people who originated in the Mediterranean basin, but extremely rare elsewhere, predisposes to the disease. Broad beans, dried for winter use, have certainly featured down through the ages in the diet of the rural poor in Britain. They have also doubled as animal fodder, giving a high-protein friskiness to horses 'full of beans'.

The other beans marketed fresh in the pod are from the New World. The Scarlet Runner and Kidney beans originated in South America and were brought to Britain soon after the great voyages of discovery but did not become popular for many years. Scarlet Runners were grown as ornamentals, and were appreciated as food only in the last century. Kidney beans in their great variety of form and name — French, Bobby, Snap and Dwarf beans — are mollycoddled in greenhouses for profitable early returns in many market garden districts as well as being grown on a field scale for freezing in the eastern counties. Runners are a speciality of the Evesham district, and bring a bright dash of colour to high summer as they wrap around their long cane tripods with scarlet curtsies all along the way. The seeds are often planted under perforated polythene, and the leaders pulled through a hole, to protect against late frosts. If the leading shoots are nipped off, the beans produce bushes bearing what are then called 'ground' beans, a week or two before the main crop. The latter are grown up canes, laboriously tied up every summer, to produce long, straight 'stick' beans which commandeer the market as soon as they are ready.

South America is also the origin of maize, first cultivated by the forerunners of the Inca civilization around 4000 BC. The old English word *corne* described the staple grain of a country, so that for the English it means wheat and for the Scottish, oats. For the Americans corn means maize. Vast quantities of high-starch varieties are grown there for animal feed. 'Flour' types are used for making corn flour, cornmeal and the familiar breakfast cornflakes. In Mexico they provide the staple *tortillas*. There is a very hard maize which bursts upon heating to give us popcorn. And there are varieties bred specially for a high sugar content called sweetcorn, for which Britons have recently developed an avid taste as a fresh vegetable eaten on the cob. About 35 million of these cobs are now grown each year, mostly by specialist growers in the southeast and on the Isle of Wight where sunlight required to ripen the ears is more dependable. But several growers in Evesham have tried out the crop.

Radish is our quickest-growing salad, needing only twenty to thirty days from sowing to market, and serves as a useful 'catch crop', either in greenhouses or outside, between major ventures. Radishes were listed on a quartermaster's tablet accounting for the provisions of the slaves who built the Great Pyramid of Cheops, and they are mentioned in a Chinese work called the Rhya, published in 1100 BC. They are now grown in virtually every part of the world and the range of varieties is enormous. There is the white Icicle, and the winter Round Black Spanish and China Rose. The jumbo-sized Daikon of Japan was traditionally carved into an intricate lattice to cover baked fish. There is a bitter, white, Japanese-type called Mooli which is about 10 inches long like a thin parsnip and is usually cooked like a root. And there is a mild white type about the size and shape of a carrot called Rettich which has been pioneered on British markets by the Dutch. Evesham growers tend to stick with French Breakfast and Cherry Belle.

The more exotic salad plants are most frequently found on smallholdings which specialize in herbs. They include various red-leaved lettuces and endive, and those salad plants which were gathered from the wild in rural districts well into this century, such as sorrel, Good King Henry and corn salad. I understand that there is demand for these salads from some restaurants, but the quantities entering the trade are minuscule. Not so with the more common herbs. The Vale is the main centre of parsley production in Britain, and although the market is usually finely balanced between adequate supply and glut, protected crops can attract high prices in the winter and spring. When prices reached £10 for a 5-lb boat during a recent season, the crop was known on the markets as 'green gold'. Sage, thyme, mint and rosemary are also grown on a relatively large scale. Grown for drying, these herbs can usually be harvested at a time when other work on the land is light. Partly dried on site, they can then be kiln-dried in a processor such as the one operated by the Littleton and Badsey Growers Co-operative. Out of 8 tons, fed leaf, stalk and all into the kiln, some 3½ tons of separated, flaked product will eventually emerge, ready for sale to the packers.

I am not aware of anyone drying and processing the Capsicum and chillis which are the currently favoured alternative crops of so many greenhouses in the Vale. Capsicums are also called peppers but should not be confused with the pepper which is the most important spice in world trade, the fruit of an Indian vine much used for flavouring savoury food. Black pepper is made from the sun-dried unripe peppercorns, white pepper from ripe peppercorns which are soaked and rubbed to remove their outer coating. The capsicum 'peppers', rather, are a group of plants native to tropical America and the West Indies, and there is a large range of varieties. Some of these produce the small, red chilli peppers, also sold in their slightly unripe green condition. Some varieties of chillis can be extremely pungent, and are one of the essential ingredients of most curry powders, chilli powder and tabasco sauce. Cayenne pepper is made from dried, powdered chillis. The mild sweet peppers which are rapidly becoming a standard ingredient of our salads were brought to Europe by Columbus and have been grown in Hungary and Poland for over three hundred years. The Hungarians have given us their name 'paprika' for the fully ripe red fruit which is dried and ground to make a mild spice. In the Evesham hothouses, however, the peppers are usually picked green for quicker returns and to encourage the plants to set more fruit. The crop is a relatively easy one to grow, although considerable research is continuing to determine the most suitable conditions for the various types.

Aubergines, like chillis, Scarlet Runners, potatoes and many other plants, are actually perennials in their native environment. In the case of aubergines this is tropical Asia, where variously coloured fruits grow on dense, spreading herbaceous bushes up to eight feet high. In the greenhouses of Evesham, growers have found that they can be grown quite profitably on trim plants about four feet high in neat rows, as annuals. But the glossy and deep-purple-skinned fruits which droop underneath the foliage like enormous enchanted tear-drops must look to many in the older generation like the epitome of everything unfamiliar and seductive-looking in nature that their mothers warned them not to touch. Introduced to Europe via Northern Africa and Moorish Spain in the Middle Ages, they were first called Mad Apples in Britain, in the belief that if eaten they would cause insanity. In fact, of course, the aubergine, or eggplant, is quite bland, tending to adopt the flavour of whatever it is cooked with. There is a restaurant in the Middle East which boasts that it can serve aubergine in over a thousand ways. And in the Balkans the vegetable has an everyday role in the diet rather as the potato, a member of the same family, does in Britain. With an estimated total area of some 15 acres grown in Britain, we have some way to go. If we wish to.

The Vale, like all market garden districts, has a good share of Pick-Your-Own farms and, although most of them concentrate on the most popular soft fruits, some do plan for a wide range of vegetables to be available at the same time. In either case many PYO growers acknowledge a keen interest from their customers, and help to develop it with information about the different varieties they have on offer, each with its own merits. This furthers a sense of adventure, and discrimination, which growers are increasingly aware of in the consumer, even when several steps removed along the distribution chain.

In addition to the PYO operations, several soft fruits are grown on a commercial scale for conventional trade, especially in the west of the Vale around Pershore. Most important of these are gooseberries and currants, both species of the genus Ribes which once grew wild over much of Europe. Gooseberries earned their name from the Tudor custom of cooking them into sauce to accompany roast goose, displacing a rival sauce made from the red berberries once common in British woodland. Before that, gooseberries were known as feaberries or fayberries, either because their juice was used as a herbal treatment for fevers, or because fairies in danger were supposed to seek protection in the middle of their prickly bushes. Botanists suspect that they were first domesticated by cottagers who selected better samples from the wild and cultivated them in their gardens, although improved bushes were introduced from France to the royal gardens of Edward I, and again during the revival of fruit-growing in the reign of Henry VIII. Their cultivation became widespread towards the end of the eighteenth century and Britain gained an international reputation as the home of the fruit. The fruit inspired a cult following among amateur gardeners, especially in industrial Lancashire, where dozens of gooseberry clubs flourished — usually based at inns or pubs whose landlords offered trophies for prize berries. By 1831 the Horticultural Society listed 360 varieties of gooseberry, with names like Warrington, Lancashire Lad, Cheshire Lass, Ironmonger, Postman, Dan's Mistake and King of Trumps. The declining demand has meant that little development has taken place this century, and the Careless, Leveller and Whinham's Industry most widely grown today all date from the 1850s. This has been possible because gooseberries are not greatly weakened by virus diseases, although very recent work to achieve resistance to diseases such as mildew has produced Invicta, and others are promised.

Currants found their way into cottagers' gardens even later than gooseberries, perhaps because the wild forms were so small and unpromising. When writers first took note of them, in the sixteenth century, some mistook their fruit for the dried currants used in confectionery which are obtained by drying the tiny purple Corinth grape. The word 'currant' stuck, although one writer insisted on calling blackcurrants "bastard Corinthes". Blackcurrants had a poor press for many years. In 1633 they were described as "of a stinking and somewhat loathing savour", whereas the distinct species of redcurrant, and its pigment-less 'white' strains were increasingly popular and quite quickly improved. Redcurrants were actually forced in heated greenhouse-like structures, and covered against frost at the end of the season, to extend their availability. It was not until the much higher vitamin C content of blackcurrants was appreciated, earlier this century, that they increased in popularity, with a corresponding decline in favour of the red. We now grow some 16,000 tons of the black against 1,000 of the red and white. Many of the blackcurrants are grown in Kent and Norfolk for mechanical harvesting and juice production, but there is still a vibrant trade in the fresh fruit from the West Midlands.

The blackberry is another soft fruit which has been improved in size and texture by cultivation, some 4,000 tons of varieties like Bedford Giant and Himalaya Giant being grown each year on wire supports, rather like raspberries. Blackberries were not brought into gardens until early in the last century, but rather because they were so good in the wild than because they showed so little promise. The wild blackberries which we still consider worthy of our attention today are probably very similar to those which were widely available to our ancient ancestors. The seeds of blackberries have been discovered in inter-glacial deposits, and at Iron Age, Bronze Age and Roman sites in various parts of Britain.

There are regular appearances of new hybrid berries raised by crossing different species of Rubus such as blackberry and raspberry . . . the sunberry, tayberry and tummelberry, for example, following the earlier generation of Veitchberry and Phenomenalberry. . . . But there is a quite different berry which has been tipped as the one with the brightest future. This is the blueberry, a native of eastern North America related to the small bilberry which grows wild on low shrubs on our heaths and moorland. The highbush blueberry is a bushy shrub, pruned to a height of eight feet, which thrives on moist, acid, peaty soils. Each established bush can produce about 10 lb. of juicy, highly flavoured berries up to half an inch in diameter, each year for twenty or thirty years. It was first grown in Britain in the 1950s on a farm in Dorset which has a naturally acid soil. Then in 1980 Mervyn Needham from Kidderminster, some twenty miles northwest of Evesham, became fascinated by the crop while on a study tour of New York State organized by the Pick-Your-Own Association. On his return he lowered the pH on a ¼ acre of his land by giving it heavy applications of sulphur, and planted 360 bushes which he obtained as cuttings from the grower in Dorset. "Most people wouldn't consider doing something like that, and it would be quite hard on heavier soils than mine. We have to net the berries or the birds will take every one. And we have to irrigate constantly through the season, even with a thick sawdust mulch to keep the moisture in the soil. So it's an expensive crop to establish. My wife says it's just a hobby. But now, after four years, we are starting to see profits, and the future looks very good. The bushes have a pretty white blossom with a lovely red leaf in autumn, and it's a delicious fruit to eat. The berries hold well on the bush for ten days once ripe, and another ten days after picking. I wrote to the *Grower* magazine asking if any other growers were trying blueberries, or interested in the idea, and received over thirty answers."

Among all the adventurous growers in the Vale and its environs, however, I could find none growing commercial quantities of artichokes. I mean here not the small knobbly tubers of the Jerusalem Artichoke — a relative of the sunflower brought back by the French from Canada and once grown widely as windbreaks and game cover in the eastern counties — but the globe artichoke much loved in Italy. Named after an Italian dialect word for pinecone, these young flower heads are borne in profusion on perennial plants which are easily propagated from suckers which grow from the base. The plant was probably first cultivated in Italy, and was certainly appreciated by the Romans. It was brought to Britain during Henry VIII's reign and was widely grown in private gardens. Artichokes were described in cookery and gardening books for the next two hundred years, but then lost their popularity and were mentioned only in medical texts — as a good tonic for the nerves. Today over half the world's crop is grown and eaten in Italy, but we are learning again that we are Europeans as well as Britons, and the artichoke is back in our diet, courtesy of considerable imports from Italy, Brittany and Spain. With numerous families from Italy already cropping land in the Vale, there is surely no need for study tours abroad. Perhaps the small patch of artichokes I saw on the Italians' holding when I started my tour of the Vale near Bredon Hill was for their own table. Or perhaps it was a first tentative effort at establishing yet another crop in our market gardens.

16

Plum Good — Damsons in the Lyth Valley

HARRY WAS MY favourite, among the many lorry drivers I rode with as a boy. Harry taught me how to tie dollies to secure the crates of Chilean onions we collected from the Alexandria docks in Liverpool. He showed me how to sheet up a load of potatoes. He knew the best transport cafés and roadside caravans, where he would buy me tea in pint mugs and bacon sandwiches whose whole slices of thin white bread oozed brown sauce from within, and bore large brown fingerprints without. But above all, on the Tuesday and Thursday run-up to the hospitals in Kendal and Windermere, it was Harry who let me drive his lorry down the long, deserted private driveways. It was the most casual of apprenticeships, and for two or three weeks in mid-September we would return from this Lake District run with one of the most casual of British harvests.

Driving west from Kendal, Harry would shout some lewd tale at me over the rattling hump of the engine. The vehicle filled the narrow lane, and even from my high-perched seat amidst the coils of rope, I could not see over the banks on either side of us. The road climbed the limestone barrows with many twists and turns and then dropped steeply by Underbarrow Scar and across Underbarrow Pool to the parish of Crosthwaite and Lyth. I remember dusk always closing around us as we dropped down to Crosthwaite at the end of a long day.

The two valleys which run their brief courses of some dozen miles from Crosthwaite to the head of Morecambe Bay — the River Gilpin and the River Winster each making its own way around Whitbarrow — are seldom included in the guide-books of the Lake District. I cannot imagine a feature excluded in a nicer fashion than was the Winster Valley in W. Pearson's little book of 1850: "It is not our business to describe this obscure vale — its sweet natural woods, clothing the gigantic side of Cartmel fell with its green meadows and sloping fields, its simple farm houses planted in all kinds of snug

corners. We will leave it, like modest merit in its own blessed
retirement."

Harry made his way according to some sure instinct, like a
migrating bird, to the particular farmhouses in their snug corners
which he had last visited a year ago. Yes, there would be the smell of
woodsmoke from the farmhouse chimney. And in the centre of the
cobbled yard, a pile of damsons waiting for us. They were weighed to
the score, which in the past in the Lyth Valley meant 21 lb. They
would be packed loose in the same odd collection of containers which
Harry had brought back with him on the final trip of the previous year
— Guernsey tomato trays; the sturdy little boxes of the Lancashire
tomato growers, stamped 2s. 6d. deposit; a few flat strawberry trays
bearing the names of various farmers in Wisbech; some cardboard
boxes which were already damp, waiting for the most propitious
moment to break and send their contents rolling. . . . And in front of
this unkempt harvest awaiting its bouncing ride to the big city on the
back of Harry's lorry, a sheepdog kept a watchful, curious eye. It
seemed to take him only a second to recognize the smell of Harry's
pants from the year before.

Damsons took their name from the city of Damascus, so the story
goes. The crusaders brought them back in the Middle Ages, as a
source of red dye for the Kendal wool trade.

At the end of the thirteenth century there were an estimated 8
million sheep in England, which meant there were several times more
sheep than people. Mutton and sheep's cheese formed a significant
part of the national diet (Walter of Henley estimated that 20 ewes
would make milk for 4 pints of butter and 250 lb. of cheese every
week) but in terms of trade, wool was by far the most valuable
product. England was the chief source of raw wool for all Europe and
was in the process of developing its own manufacturing industry. The
crusaders came back almost penniless, but some did bring with them
new breeds of sheep from Syria, and all had been exposed to more
splendid and ostentatious civilizations abroad. The damson provides
an excellent rich dye, and grows easily from its stone. It would take up
little room in a saddle-bag.

In the local history section of Kendal library there is to be found
Gate's Shepherds Guide, *Hodgson's Shepherds Guide*, *Lamb's Shepherds
Guide* and several others. A wool trade there is, and for centuries has
been. But with the help of two librarians I could find no mention of the
medieval damson, and of no local gent who travelled east to fight the
Saracen.

"They were never called damsons here, anyway" says Desmond Holmes. "Local people still call them simply 'the plums'. So it makes you wonder about the Damascus connection."

Among other things, Desmond Holmes farms 100 acres with a herd of beef cows and calves. He once attended evening classes in the methodology of medieval research. "There is an effigy of a crusading De Beetham in Heversham Parish Church," he told me, "and at Levens Hall there are Baggot goats which came from Syria. But no one really knows about the damsons." He says this rather wistfully, as though he would dearly like to take on the research himself, and perhaps even protectively: the definitive account awaits his pen.

"After the Enclosure Act of 1803 farmers used damson trees to mark that they owned their own fields. Until then the valley had been an open common grazed by cattle and sheep. They kept on planting damsons along the walls and hedges where they didn't interfere with the pasture. They're still there, marking the fields. And they look very attractive, especially in the spring."

I talked to Desmond Holmes at Whitebeck Farm, which his family has owned since the eighteenth century. For the last twenty years Whitebeck has been let to tenants but now Desmond is moving back here from Kendal. He was in the process of putting in a damp course, replacing window frames, and replastering. In the former cowshed built on to the house he showed me his collection of old farm implements: simple board ploughs, drainage tools, spades for cutting peat. . . . And on the slope above the house the "new orchard", the last intake of his land given to the damson. It was planted by his great grandfather Taylor in 1879.

Early this century there were 400 orchards in the Lyth Valley. 'The plums' were carried by horse and cart to Milnthorpe or Sandside stations and then by rail to the big city markets. Fifty or sixty carts would load at Sandside each day, according to one report. The damsons were sought after by the largest commercial jam makers in the country. They were even shipped to France.

During both World Wars women of the Land Army were stationed in the valley. Hundreds of tons of damsons were harvested in those years, together with large quantities of hedgerow blackberries and rosehips for a country starved of vitamin C. Italian prisoners of war from Bela River prison camp at the mouth of the valley also worked on the farms of Crosthwaite and Lyth. They got on well with the farmers and after the war several of them chose sure employment in the valley rather than an uncertain future in the impoverished farms of southern Italy from whence they came. Forty years later I found the last Italian picking damsons in the Lyth Valley, and he was due to retire the following Christmas.

Salvatore Citino hauled his ladder round the other side of a twenty-foot-tall damson tree from Desmond Holmes and then climbed up it with his bucket. He wore baggy brown pants with broad canvas braces over his brown shirt, and a beret. And he spoke with a thick Italian accent: "Oh yes, yes, it was lovely when we got here, when we got to Bela. But the journey, that was bad. It took fifty-five days, round the Cape. Fifty-five days, to avoid the submarine."

"He's the man who let us take Tobruk," Desmond shouted with a laugh from within the tree.

"But they treat us good here. In '48 I come back with my wife to stay." He waxed nostalgic for a moment about a youth spent picking tomatoes and olives, grapes, walnuts and lemons. "But there was no market. You couldn't sell it anywhere." He has been back to Italy to visit relatives twice. But he likes the Lake District. He says he even likes the rain.

"I planted this tree," says Salvatore. "About thirty years ago." We were in the Long Garth orchard below the house.

"You can see we don't believe in pruning," Desmond joked. But they had picked over seven score of plums from less than half of the tree.

The Lyth Valley was carved by a glacier, and the underlying Silurian slaty rock is covered with glacial deposits or 'samel' which are quite deep and workable. On the eastern edge, Underbarrow Pool flows under the steep limestone scar which protects the valley from cold winds and snows. 'Pool' is the old local word for river, and the River Gilpin which flows down the western side of the valley just below Whitebeck, was once called Lyth Pool. 'Lyth' derives from the Scandinavian for slope, as the valley on this side slopes gently up the back of Whitbarrow. It is the physical shape and position of the valley which creates a very local climate which in turn has created a damson crop with such a widespread reputation. There is nothing higher than a hedgerow between the valley floor and Morecambe Bay, and it is the mild, tempering effect of the bay which brings the blossom early and gives the fruit such a good start. Beekeepers bring their bees here especially for that reason, moving them next on to the lime and heather. "Get up to Crook or Winster, less than three miles to the north of us," Desmond said, "and you can hardly call them plums."

On the roadside Sabina May Holmes, Desmond's mother, had set up a trestle table and her old-fashioned market scales. In between the damsons she set a tray of mushrooms gathered from the fields that morning, and boxes of apples and pears. There are Russets, Millers, Scotch Bridget, Baron Worseley, Rank Thorn, Hazel, Doctor Joules, Fertility, and many other unnamed varieties of apples and pears in the small Long Garth orchard.

"It's not really a commercial proposition," Desmond explained, as he tipped another bucket of plums into wooden trays. "Not with labour the way it is. Not at 12 pence a pound wholesale, or 18 pence at the farm gate. But I hate to see the fruit wasted so I get out here as often as I can at this time of year and pick as much as I can. If it rains heavily now the plums will burst and that will be that." He talked a little about the politics of farming and revealed that the present Minister for Agriculture was the Member of Parliament for the Kendal district. "You go and ask Michael Jopling about the future of the Lyth Valley damsons," he told me jokingly. "That will shut him up for a moment or two."

No new damsons have been planted at Whitebeck for the past twenty-five years. "A damson can fruit well for a hundred years. We used to replace trees when they died, or when they were past it. We'd get the new trees out of the hedgerows mostly. They grow true from stones. They take about fifteen years that way, to produce fruit. You'd have to cut the pasture for hay until the trees established themselves. But we don't replant now. We graze stirks under the orchards." And then he's away on another historical tangent. "A stirk is the local name for a calf in its first year . . . from Strickland parish, or 'Stirkland' as it is in the Domesday Book. . . ."

Nationwide, the damson forms only a small part of our plum harvest, yielding around 2,000 tons based on 5-year averages, compared to 30,000 for all other varieties. But its inability to command prices high enough to pay for the labour of picking it is typical. In addition to this, other commercial plums are highly vulnerable to disease and to adverse weather — frost in the spring and too much rain at harvest time, which leads to rapid spoilage — and there is often a huge fluctuation in the size of the crop from year to year. In all of our chief plum-growing districts — Kent, the Vale of Evesham and East Anglia — orchards are being progressively grubbed up and put to more profitable use. The area devoted to Victorias was down 5 per cent last year, the late varieties fell 10 per cent, and the area down to Pershore Yellow Egg, no longer in great demand at canneries, fell 30 per cent. This decline has lasted for over ten years and is likely to continue. The £54 per ton offered to plum growers by processors last year was described by a member of the West Midlands Fruit Growers

Association as the lowest offer since 1936. The growers who remained, he said, were producing the crop only because of the expense of grubbing up their orchards.

In 1984, 30 of these Evesham growers, who produce over three-quarters of the district's plums, formed a marketing consortium. They aim to control quality and co-ordinate marketing, which means never allowing the market to become flooded. They have a brand name, Vale Plums, and PR advice from Food from Britain. A giant multiple retail company contracted to take a quarter of the group's plums in their first year — the first time English plums have ever been distributed nationally by a supermarket.

So perhaps our plums will go the way of our cherries, becoming a high-quality luxury item. Technology will be applied to this end. Fruit size can aleady be improved, for example, by blossom-thinning using high-volume chemicals (in the early morning or late evening to avoid killing bees) or by fruitlet-thinning using growth-regulating hormones. We shall lose the rough and ready, and we shall also lose our choice. The Vale Plum scheme, for example, only extends to Victorias, damsons and the late season Marjorie Seedlings.

The great virtue of choice in the past was that while the season for any one variety could be as short as two weeks, the earlies such as Rivers, Prolifics and Czars started in mid-August and the beautiful if bitter Wyedales appeared, invariably dusted with bloom, as late as November. We could have fresh stewing fruit, straight from the trees, throughout this period. My personal favourites include the oval Purple Pershores which fall easily from the stones; the greengage brought over from the continent by Sir Thomas Gage in 1725 and the Warwickshire Drooper, both among the best of our pie and pudding plums; and the late, full-fleshed Monarch. There are many, many more. A concerned gardening club at a school in Evesham has begun a living plum tree museum by planting some of these varieties. And no doubt many of them will survive in private gardens, just as the ancient bullace plums, and the wild sloes and cherry plums which by their interbreeding through thousands of years have given us such a variety in cultivation, can still be found in European woods and hedges.

The plum which our commercial growers now seek to emulate is the Japanese type, grown widely in the United States and South Africa, but flowering so early that it cannot be grown in northern Europe. These plums, such as the Santa Rosa, are vastly inferior to our native plums in eating quality, but they win hands down on the ascendant values of size and showiness. In seventh-century Japan when only plants of ritual significance were allowed in gardens, these plums were definitely in. Ladies made special dresses for ceremonial moonlight visits to the plum-blossom.

Following afternoon tea with damson cheese at a Crosthwaite Inn, I drank damson brandy with Brian Walling. He lives in an early eighteenth-century cottage called West View on the other side of the valley from Whitebeck and he is a breeder of pedigree Herefords with a national reputation. He has 10 acres of damsons in three orchards and in a good year crops 33 tons of plums. His damson drink has a kick like a prize bull. You can make a powerful brandy-type liquor by putting damson wine in the freezer and picking out the lumps of ice from the water which freezes. The flavour and the alcohol undergo a remarkable concentration.

"We used to have 25 ladders and 40 pickers," Walling told me. "But nobody wants to pick now. They used to come from Kendal. But there's laws against children picking and the tradition has died. You have to grow up picking. You have to learn the way of it young." The young are leaving Crosthwaite. "There were 70 at the village school when I was a lad, and now there are less than 30. A good picker can earn £100 to £130 a week, but nobody seems to want the work. The teenagers all want their weekends away."

In recent years Walling has driven his plums on his own truck to the wholesale markets of Manchester, Newcastle and Glasgow, where Lyth Valley damsons often command a premium over those from Kent or Evesham. But this year he can find no labour at all. He is considering placing an advert in *The Westmorland Gazette* for Pick-Your-Own but he is not optimistic that there will be much response. "They'll only pick the easy ones," he says despondently. "And they'll do too much damage with the ladders." He means damage to the trees, not to themselves.

"We always used to replace dead trees with suckers from the base of others. And there's pruning to be done, dead wood to be sawn out. It's not worth it any longer. It's a shame, because they're damned good plums." I walked up to his largest orchard, on a southwest facing slope above his farmyard. There were piles of brush from recent prunings and evidence that a chainsaw had been at work. And as yet there were no large gaps among the trees. But Walling reiterated his despair. A crop which at the turn of the century had paid the rent for most farms in the valley, and is still a nice perk at the tail-end of the year for some, has become a nuisance for others. Potentially good grazing is reduced to poor grazing by the presence of trees whose excellent harvest has gone the way of teatime, and whose days are numbered.

Walling continued in the same mood: "The valley used to be a tight community, with families that had been here for hundreds of years, and all the youngsters going to school together. But it's changing now. All the picturesque homes are being sold to these retired people." I thought he said "artsy types" and then possibly "barristers", but when asked again he said "environmentalists". . . . "They come here and start telling you what to do and what not to do."

On my way out of the valley I discovered that a pumping station had been built at the southern end, and that in future the valley floor would be progressively drier. During the last war when it was compulsory to grow corn, flooding made it hard to get the grain off in the autumn. Now, I discovered, environmentalists were concerned that farmers might plough up the whole valley. And they feared the possible arrival of a large glasshouse industry.

Damson dene it isn't. Nor could I leave the Lyth Valley, "like modest merit, in its own blessed retirement".

17

Tender Hearts — Celery off the Isle of Ely

A FEW MILES from Ely, on ground which once formed the bed of Soham Mere, I tried to conjure up a former scene: of William the Conqueror's heavily armoured soldiery floundering among dense shrubs and tall bog reeds in their unhappy pursuit of that Anglo-Saxon hero and early exponent of guerrilla warfare, Hereward the Wake. It was not easy. The ground beneath my feet was firm and dry and for hundreds of yards in all directions nothing grew above knee-height. Or below knee-height. The vegetation all round me was exactly at knee height. I was in the middle of a laser-levelled, precision-planted celery field. The heads grew nine inches apart in rows of six to each six-foot bed. That gives 50,000 head of celery per acre, and it was a 30 acre field. We grow some 120 million heads of celery each year in Britain, and over half of them come from the flat fenland fields which encircle the small outcrop of rock at Ely.

Celery has always been a bog plant, growing wild in moist areas, especially near the sea, from Britain through central and southern Europe to western Asia. Its ancient form was very similar, if not identical, to the bitter-tasting herb called smallage which is still occasionally grown in gardens for flavouring. Our word for the crop derives from the Greek *selinon*, meaning parsley, perhaps because for centuries celery was used almost exclusively as a herb for seasoning. The first record of celery being used as a food was in France in 1623, and from that date horticulturalists in France and Italy began selecting strains with a longer leaf stem and milder flavour. But it was only as recently as some two hundred years ago that they succeeded in producing celery sticks which resembled those which we enjoy today. The practice of drawing up the earth around the growing plants was a decisive factor in producing long, mild-flavoured stalks which were crisp and succulent with a minimum of stringiness. It also had the effect of blanching the stems, and special red willows were grown to provide splicing twine which would best set off the whiteness for sale,

although the Victorians also favoured red strains of celery which
'blanched' to a delicate shell pink. Fenland soils were well suited to this
blanching technique, but the process was highly labour-intensive.
And the peat was persistent, despite the considerable labour of those
who practised the specialized trade of celery-washer. Continuing
selection, however, has given us the American green types, and self-
blanching varieties which are easier to grow, more attractive to
market, and of high eating quality, without requiring earthing-up.

Jeremy Harwood is the production manager at G. S. Shropshire and
Sons, who describe themselves as 'Growers and Packers'. This firm is
perhaps the foremost celery producer in the country, and one of the
first to wash and pre-pack celery for the supermarkets. My moment of
contemplation in the flat, verdant landscape of their celery fields was
the briefest part of a tour which concentrated far more on the
industrial-like processing of the crop between harvest and sale. And
the eminence of Jeremy Harwood's role in the company confirmed the
equal importance of the 'packers' with the 'growers' in the firm's
operation.

"The celery is brought to the packhouse in plastic crates on tractor-
trailers and is tipped out on to moving belts on one of eight lines.
There are five women on each team and they are paid piece-rates per
team. The first two line the celery up for the automatic cutters which
trim off the top and bottom. The waste drops down into a water flume
which carries it out to the rubbish heap. Then the heads are retrimmed
by hand and the outer stalks picked off. This material is dropped into
another flume and transported automatically into a processing unit for
freezing or soup, with whole small hearts selected into a heart pack.
We used to can about 80,000 cans of celery hearts a year, but now they
go into special packs for the supermarkets. The prime heads are then
carried off in the continuously moving four-tier chain of baskets and
pass through a series of high-pressure jets which wash and then drain
and then wash and then drain until they come round for pre-packing
and grading. Each packer has up to twelve different bags to fill, with
four grades, and is trained to recognize these by eye. So a Sainsbury
head goes into a Sainsbury bag and a Sainsbury box, and she'll be
packing numerous boxes simultaneously. She has a little abacus above
each to keep count, and her boxes are coded with her number so that if
we have a problem with quality control we can trace where it comes
from. The full boxes slide down the rollers on to pallets and when a
pallet is full it goes into the vacuum cooler for thirty to forty-five
minutes to take the heat out. We get it down to 4°C and then move the

pallet into the cold store, where the temperature is kept about the same. From there it goes into a refrigerated lorry and is usually delivered the same day."

This celery washing and pre-packing line, which can take 120,000 sticks per day, was personally designed by Guy Shropshire, as were various other devices on the farm. The Shropshires farmed originally in Shropshire, in an area once renowned for carrot production. Guy was set up by his father as a tenant on a 250-acre mixed farm at the age of twenty-two. Three years later, in 1952, he bought 330-acre Fordey Farm and then in 1957 neighbouring Hainey Farm, near the village of Barway three miles south of Ely. On this fenland he concentrated on vegetables, with celery and carrots in particular. Later acquisitions brought his total acreage to 2,700 with a range of different soils which enable him to grow a diversity of specialist vegetable crops which can be efficiently integrated within a year-round packing and marketing organization. On land situated in the triangle where the river Cam meets the Great Ouse, a few miles south of Barway, for example, centuries of flooding have given a super-fertile layering of peat and silt bands. There is some lighter fen and sandy soil at farms just inside Norfolk to the north, and blackland fen with islands of clay and greensand around Barway. Guy's two sons are largely responsible for managing this land, with John basically running the Cambridgeshire farms and Peter the Norfolk ones. Guy Shropshire is still very much the leader, however. "He pops into my office virtually every day, to keep me on my toes," says Jeremy Harwood.

"My roots are in this area and I've known the Shropshire family a long time. I was offered a job here when I left school but I didn't take it because I wanted to get some experience of other things. I worked for Cadbury-Schweppes at Cambridge and then Bournville. Then Mr Shropshire offered me a job as packhouse manager. So I started here thirteen years ago when we had 30 people working about six months of the year packing just celery. We've added extra items each year so that now we have 230 people packing fifteen to twenty different crops through the year. We do a great deal of iceberg lettuce and Little Gem, which is a Cos-type. We pack a lot of leeks, out-of-season new potatoes, and Chinese leaves. We grow celeriac and about 10 acres of fennel . . . there's a big range. . . . We trialled twelve new products this year alone, including red lettuces, and have isolated two of them to pursue further. That's in addition to a large acreage of onions and main crop potatoes, and sugar beet and corn.

"We grow about 350 acres of celery, which includes about 70 acres which is earthed up for a Christmas dirty pack. We start lifting the narrow row crop in late June and keep going with that, planted in rotation, up to the end of November. Then the wide-row, earthed-up crop takes us to Christmas." Celery is a biennial plant which, if not harvested in the first year, would cease to grow at temperatures near freezing and then send up a three-to-five-foot-long flower stem the following season. This would bear a large white flower and eventually a large number of seeds. These are among the smallest seeds of cultivated plants and are occasionally grown and sold for seasoning. "The problem with celery is not so much that it's not frost-hardy, but the fact that when you get frost the outer membrane shrinks and the stick becomes translucent. If you get too much frost the thing starts to rot, but basically the stick becomes untidy. It may be edible but it doesn't conform to EEC standards and becomes unsaleable. So to give us year-round capacity we are involved in growing celery in Spain, which takes us through until the end of April. Then we market Guernsey and UK glasshouse celery from associates such as Home Grown Salads who are within our growers' co-operative, G. S. Growers Ltd.

"It would be very nice to find a frost-hardy variety, but we don't think we'll succeed. Having said that, we've got a trial this year of 26 new varieties for taking us through into the 1990s. At the moment we're growing mainly Celebrity and Jason, which are offshoots of Lathom Self-Blanching which was the original self-blancher first grown about twenty years ago. We're also growing some American Green celery. Flavour is the main thing we're looking for now, and American Green has more flavour and a higher vitamin content, keeps better, and is less stringy. The stringiness problem is complicated because the strings you're trying to avoid are actually the tissues which carry water and nutrients to and from the leaves. We grow American Green for export and processing, but although the British public will buy it from Israel and elsewhere out-of-season, they won't buy it in the main season when white is available. As for the dirty celery, our main varieties are Multipack and Ely White, which is a locally raised one. The most familiar is Fenlander, which was raised on one of our farms before we bought it, but we grow very little of it now because it has degenerated . . . it's very difficult to trim and bring up to standard . . . and there is so little grown now that there is not much effort going into sorting the seed out."

The celery is raised in peat blocks under glass and planted out from late April onwards. Celery is very slow to germinate, and this can be facilitated by hormone treatment of the seed. Weeds are a great problem on peat soils, and herbicides are applied to the land before planting and up to five times after planting. Because of the constant irrigation and high humidity which the growing crop needs, there can be problems with fungi requiring fungicide sprays, and pesticides are applied to protect against celery leaf-miner. But Jeremy Harwood says that there is an awareness at Shropshires of public concern about excessive use of chemicals, and all applications are kept to the minimum required. A major retailer recently tested various samples of their produce, he says, and was unable to detect any residues. Some of the celery is also sprayed with the hormone gibberellic acid, which encourages a longer stem growth, and can be used to adjust the rate at which the crop matures, to ensure a smoother harvesting programme.

"This is the first year that all the celery fields have been laser-levelled before planting, and from my point of view it has paid hands down. The machine cost £120,000 and consists of a four-wheel-drive tractor towing what looks like two dozen blades behind. In fact they are laser-controlled to scoop up soil where it's high and drop it down where it's low, to give a true horizontal. This gives us a level water-table throughout, so we can control the irrigation precisely, and this has given us a more uniform, better quality crop than we've ever had before. Irrigation pipes are laid as we plant the crop, with standpipes where needed. We can take water from the dykes round the fields, and we can control the level in the dykes by taking water from the river, which is actually higher since the peat has shrunk over the years, or pumping it up into the river. We keep the water-table about thirty inches below the surface." I asked if this were checked automatically by some robotic device and he laughed: "The manager goes out and digs a hole with a spade and measures it."

There is another machine, however, which plays an important role in the production of fenland celery, and that is the Shropshire straw-planter. The really black peat soil is reserved for leeks and potatoes, with the special advantage of being quick-drying and therefore workable all year round. Celery is best grown on a mixture of peat and silt, but this soil can still blow in a strong wind. "Ten years ago Guy Shropshire invented a machine which lays straw on the soil after drilling and then cuts it in, so that you have a mini-hedge standing nine to twelve inches tall every eighty inches or so across the field. This gives you a mini-climate between the rows of straw so you get a slightly earlier growth and when the wind blows the surface soil it hits the straw and falls down. So you prevent soil erosion and the terrible damage that high-speed particles can cause on young plants. We used to lose 25 per cent of the crop sometimes, after a heavy blow, but now we never have a serious crop loss."

When the celery is ready for harvesting, a gangmaster is contacted. This system of organizing labour has been widespread in East Anglia for many years. The gangmaster contracts to supply the labour required for a fee, but these days works with the farmer in the supervision of labour. A tractor-mounted machine trims the leaf on the standing celery down to the required height using a device rather like a large rotary mower, and then undercuts it beneath the soil and gently lifts, and lays it on the surface. Celery is a very brittle crop which calls for a gentle touch. The gang of men follows in the wake of the machine, amidst an aroma from the chopped débris which threatens to turn the whole of the fen back into some primordial vegetable soup, and pack the heads into plastic crates which they then lift on to a waiting trailer ready for despatch to the packhouse.

It seems like a lot of energy and effort to produce a foodstuff which gives us only two-and-a-half calories per ounce. In the old, tougher, more stringy days of celery, it was said that it actually took more energy to digest the item than was gained from doing so.

In her classic book, *Food in England*, Dorothy Hartley suggested that "celery is so good raw, crisp and fresh, with cheese and biscuits, that celery hearts should be served every lunchtime all through the winter". The modern trend, however, is for somewhat more variety. Celeriac, also known as turnip-rooted celery because of its superficial similarity to the turnip, is closely related to celery, and was developed through selection by the same southern European horticulturalists who first improved stalk celery. Its leaf stems are short and bitter-tasting, and it is the swollen, root-like base of the stem which is eaten. This can be peeled and grated for salads or sliced and boiled, jugged or added to soups and stews just like celery, with a similar flavour. Florence fennel, or *finnochio*, looks more like a head of celery, with its tightly packed swollen leaf bases, but is not closely related. Again it is often served raw in salads or boiled for a side vegetable, like celery, but it has a more aromatic flavour, resembling aniseed. Shropshires grow both of these alternative crops on a small scale.

They also grow onions.

The onion, *Allium cepa*, and many closely related species of allium, have been cultivated from very early times both as food and medicinal crops. They have been potent symbols in both folklore and religion. Onions were cultivated in the first cities of the Indus and Euphrates Valleys, and were sold by street vendors in the city of Ur at least four thousand years ago. There was a cult in Egypt which gave the onion the status of a god. Herbals from Roman times onwards have listed extract of onion as a treatment for colds and a host of other ailments, and there has been a flurry of research in recent years on the effectiveness of onions in the diet against various heart problems. The ancient Chinese civilizations used their native *Allium tuberosum* (Chinese chives) and *Allium fisulosum* (Welsh onions) both as food and medicine.

Many of the Alliums have been grown in British gardens since Roman times. They include leeks, chives, garlic, rocambole, which is a type of garlic with a coiled stem; shallots, once thought to be a separate species but now recognized as a variety of onion, whose bulbs multiply laterally; the tree onion, a perennial which bears small bulbs on its flower instead of seed, once especially appreciated for pickling onions; and the Welsh onion with thin, elongated bulbs which is not native to Wales but as I have already mentioned comes from an Anglo-Saxon word meaning foreign. But *Allium cepa* in forms very similar to the onions we know today has almost always taken pride of place, for the peasant with his bread and cheese as for the tables of the rich — "All cooks agree in this opinion, no savoury dish without an onion."

Imported onions are not a new phenomenon, for there are records of Dutch onions being shipped to Newcastle-upon-Tyne as early as 1401. During my youth I saw onions imported to the wholesale markets from dozens of different countries covering five continents, for barn-stored English onions at ambient temperatures will not keep after March, and deteriorate considerably after Christmas.

As a nation we developed a preference for the large, mild Spanish-type onions grown in hot and reasonably humid regions, and small discoloured English onions became a subject of derision on the wholesale markets, fit only for the poor end of the trade. There are varieties of large English onions grown for processing, but they have a poor flavour and will not keep. There is no Spanish-type seed suitable for English conditions — which will not bolt in our climate, for example — and these onions are also very soft which makes them less suitable for low-labour, highly mechanized handling. But in the last decade the Ministry of Agriculture has pioneered new growing and storage technology to try to win back the home market with home-produced onions. Several growers, including the Shropshires, took advantage of this impetus, and the financial aid which was available towards new facilities. In 1983, for example, a computer-controlled grading station costing £1 million was opened near Boston in Lincolnshire, pre-packing 'Lincoln County' brand onions for the 39 farmer members of a large co-operative. So that from a situation in the early Seventies when British onions met less than 30 per cent of the

British onion demand, they now serve over 60 per cent of the market.

The varieties which have helped to achieve this are largely FI hybrids raised by specialist seed companies whose parent company is often based in Holland. But just as new varieties are introduced at an astonishing rate, so we are losing the old. Lawrence Hills, Director of the Henry Doubleday Research Association, has campaigned against punitive government seed legislation which he believes has contributed to the loss of over 1,000 older vegetable varieties in the last six years. A high registration fee must now be paid before seedsmen can sell varieties which have been in common use for decades and even centuries, and are part of our national heritage. Relatively small demand means that the fee is prohibitive, and so these varieties are lost from the trade, and eventually from existence. In the case of onions he fears especially for the future of strong onions such as Up-to-Date (1890), Oakey (1880), and James Long Keeping (1834), which were bred to go with bread, cheese and ale. Or Giant Zittau, "reputed in the north to be capable of opening the garden gate with a single breath from the proud grower".

Jeremy Harwood, more used to defending his onions against the large Spanish mild, produced a single class 1 onion, almost perfectly round and with a thin, unblemished, shining golden skin. He challenged me to peel it. "Believe me," he said, "these are not sweet onions, they're plenty keen. If they are lifted wet they're less strong, but it's like anything else — a slower, harder-growing onion on heavier land will give you plenty of flavour." As I played with his onion, he continued:

"Onions were a very minor crop to start off with, but the decision having been made to go into it, Mr Shropshire did it properly. He went to Holland and investigated what was happening there. He built the proper buildings for the crop. I think we were the first in this country, for instance, to build refrigeration for the long-term storing of onions. His objective was to supply high-quality English onions for twelve months of the year, and to do this we grow what amounts to three different crops. We plant out multi-seeded soil blocks in April. These have been sown with five or six seeds per block under glass so that they are well established, with four or five inches of growth when they are transferred to the field after the dangerous frosts. They have a very good start and sort of push each other apart as they grow, forming a cluster. We can harvest these in early July to give us two months' sales until the main crop is ready in September and October. These will have been drilled straight into the field, a typical variety being Rijnsburger-Bolstora. Stored at a temperature of 31°F these will keep without sprouting until the following June. Then in August we drill Japanese onions which over-winter in the field, growing rapidly the following spring to mature in time to cover the June and July sales, and complete the calendar."

At this point he called in Fiona Brown, who is the crop technologist and one of the few female employees I had met who wore the Wellington boots of the field rather than the sterile nylon smocks of the packhouse. From her I learned that quality onions are quite a difficult crop to grow. Onions prefer a more mineral soil because grown on peat they don't keep so well, and much of the technology centres on making sure that the onions go into store in perfect condition. They also don't like acid conditions and so lime is put on — there is a long tradition in the fens of using beet factory sludge lime which is a waste product from the numerous sugarbeet processing plants in the area. They are also given a little nitrogen, and trace minerals. "Weeds are a very big problem because onions are very susceptible to herbicides. Onions go into a sterile seedbed and then you herbicide it again before they come up, leaving it as close as you dare to their emergence. Then there's a continuous herbicide pro-gramme through the crop's life, using contact herbicides which run off the onions' waxy skin. But if conditions aren't right, say the wax has been rubbed off by soil particles blowing in a strong wind, then you've got problems. We have to use hand weeding for the things like annual meadow grass which we can't kill with chemicals.

"Neck rot is not such a bad problem now because the seed is treated chemically against it. But white rot is a major headache. The fungus is viable in the soil for from fifteen to sixty years and there's some idea now that if you've got a clean field, you grow onions on it until you get white rot and then you quit. There's a fungicide effective on spring onions, but not bulb onions. There is work in the United States on biological control, introducing a fungus which will attack the white rot spores, and there's the idea of spraying extract of garlic on your field before drilling. This causes the resting bodies to germinate and then they find there are no onions there, so they all die. But this is all very ex-perimental. If you get white rot it lowers your yield, but perhaps more importantly, if you store onions which have got it, they become much more susceptible to other diseases, and it can spread through the store.

"Then there are leaf diseases like mildew and Botrytis, which can reduce yields. Slugs can be a problem in wet years, and eelworm which is hosted by wild oats and lots of other weeds. We treat with Temik at drilling, which is a pretty nasty compound, but with a lot of rain eelworm can still be a problem." I asked her if she thought these various chemicals represented any threat to the health of the consumer: "We don't use any chemicals which haven't been cleared for use. Also all these chemicals are covered by harvest limits, so that you don't use them within a certain time before harvest. I was asked by one supermarket we serve to supply a list this year of all the chemicals used on the crops, the rates we used them at, the main ingredients, and so on. And they never complained."

The onions are harvested green, two weeks after spraying with the growth suppressant maleic acid, and brought straight inside for artificial curing. The harvester consists of a topper out front which sucks up the leaf growth and cuts it off six inches from the ground. Then there is a lifter at the rear which lifts the onions across the width of the bed and leaves them in a single row. A harvester of the type used on potatoes then follows to lift the row of onions and rumble them over a riddle to remove soil and up a ramp into a trailer driven alongside. The onions are stored loose in huge sheds with ducts underneath slatted floors, which allow air to be blown through from a central chamber. First the onions are dried with ambient air, and then the temperature is increased to 90°F with a high humidity, to cure the crop.

"Curing is essentially drying, but you have to keep the humidity up to a minimum of 65 per cent, otherwise you end up by cracking the skin off. After a couple of days at that we take them right down, and keep them at 30°F or 31°F. Field-curing in Britain cannot produce the quality because if it rains you get staining, which immediately downgrades your onions. In a good year, 30 per cent of our onion crop is class 1, and we have a very high percentage of class 2, which is the grade most often sold. Marks and Spencer are the only company I know of to sell class 1 routinely, and we are one of only three companies producing them regularly." One of these other companies is Greens of Soham, who have actually been exporting onions for ten years.

Close neighbours on the fenland south of Ely, Greens and Shrop-shires are the great rival giants of celery production in Britain. They have come together in one venture, however, replacing their individual 'Hassey' and 'G's' brands with a 'Big Ben' celery export brand. With some initial assistance from the government, ongoing back-up from Food from Britain, and in conjunction with the import-export Fyffes Group, this has now become what Jeremy Harwood calls "a significant earner of foreign exchange". He had just returned from the Cologne Food Fair, where he had met current and potential customers: "During the time I was there, we had new interest from Scandinavia, more interest from Italy, and continued interest from Germany, which is our biggest export market. In a week like this we and Greens together will have sent over 4,000 boxes of celery to Germany."

18

Kings and Commoners — The Lincolnshire Potato

THE FIELD LAY within hailing distance of that wide, shallow bay called The Wash. It had been drizzling all morning and the wetness seemed to hang on the land like a shroud, removing all form and colour from the Lincolnshire landscape. The earth underfoot was soft and yielding. There was no sign of any crop, for the haulm had long since died and shrivelled, and there were few weeds. I walked towards the tractor which slowly ploughed its way back and forth across the far side of the field, the noise almost absorbed by the wet air. Then I saw the King Edwards, lifted like treasure from below and exposed in a long bare line behind the plough. The farmer stopped his machine and jumped down to show me his crop. He brushed the damp silt from a few tubers and they bloomed in his hands, and their red eyes shone. He handled them as though they were the very crown jewels which King John had lost while skirting The Wash in even wetter circumstances.

This reclaimed marshland is now among the most expensive farmland in the country. But the potato has a much humbler origin, on the cold, rugged slopes of the Andes mountains in South America, where it still grows wild.

Dr Redcliffe Salaman in his fascinating book, *The History and Social Influence of the Potato*, has given a convincing account of the role of the potato in the creation of the great civilizations of South America. There was a considerable body of opinion at one time which held that the western coast of South America may have been colonized by sea-going peoples of Polynesia. But Salaman has shown that the region was in fact settled by inhabitants of the Old World who crossed the isthmus which is now the Bering Straits, and by a long process of expanding settlements gradually pushed their way south through America. When blocked by the impenetrable and unhealthy rain forest of the Amazon basin, they crossed the Andes from the east, to reach the more hospitable plains and valleys and eventually the Peruvian coastlands.

When these early Americans emerged from the trees at great altitude they found that the low temperatures which rescued them from the diseases and other perils of the jungle also killed the manioc, maize and other food plants which they brought with them. But here they discovered wild potatoes, which grew well in the poor soil and cold temperatures. The settlers took these into cultivation and began the long process of selecting superior plants and chance hybrids. Some varieties were even resistant to frost, and have been sought after in recent times by Russian biologists eager to introduce this characteristic into new breeds for their own harsher climates. The first settlers also learned how to dry potatoes, by exposing them alternately to frost and sunlight, to make *chuno* which is still a common food in South America. Archaeologists have found evidence that the first settlements on the western coast used the potato and the coca plant native to the eastern slopes; and their pottery, which also features the potato as a powerful symbol, drew on recent memories of the Amazonian jaguar and boa to symbolize fear and evil.

Salaman claims that it was the potato which made survival possible on the highland plateaux, and allowed a subsequent descent to the inter-Andean valleys and the coast where there developed great civilizations. Some commentators have gone further, suggesting that cultivating the potato required far more social organization and discipline than did hunting and foraging for wild foods. Out of this need may have come a system of organization which would have remained intact when conditions became easier on the western descent, and which would have enabled the people to deal more effectively with the droughts which they then encountered. This could have actually inspired the highly ordered civilizations which subsequently evolved.

When the potato was finally brought to Europe it again played a significant and dramatic role in the affairs of nations. It was party to the making of Prussia and the ruin of Ireland. But here Salaman's analysis focuses not on the building of civilization, but on the exploitation of labouring people:

> If for any reason, good or bad, conscious or otherwise, it is in the interests of one economically stronger group to coerce another, then in the absence of political, legal or moral restraint, that task is enormously facilitated when the weaker group can either be persuaded or forced to adopt some simple, cheaply produced food as the mainstay of its subsistence. Experience shows that this course inevitably results in a lower standard of living. The lower that

standard, the easier is the task of exploitation and the nearer will the status of the weaker class approximate to serfdom. The potato, being the cheapest and one of the most efficient single foods man has as yet cultivated in the temperate zones, lends itself readily to the task of solving labour problems, along certain well-defined lines, in a society which, for any reason, is already stratified into social classes.

The course of events is well illustrated in eighteenth-century Great Britain: the employing class desired cheap labour; wages, in the absence of any protective mechanism, were determined in the main by the labourers' cost of subsistence; a potato diet was capable of reducing that cost to the lowest level. Hence it was to the employers' interests to urge the use of the potato on the worker, which he did directly the cost of subsistence called for an increased wage.

Salaman describes with historical detail the process by which various populations were persuaded, often against considerable resistance, to adopt the potato, and the devastating effect this had subsequently on their standard of living. In the Highlands of Scotland the potato was the instrument of an exploitation so ruthless that it resulted in the emigration of the majority of the working population.

In Ireland, where at the advent of the potato, native society was already hopelessly disintegrated, it met with no resistance and became in the shortest possible time the food of the people. In an environment poisoned by religious jealousies, undermined by political dissension, where industry was hamstrung at the dictate of an alien power, all the factors were to hand which made it inevitable that the use of the potato, cheapest of foods, would reduce the standard of living to the lowest level ever attained in Europe. After proving itself the most perfect instrument for the maintenance of poverty and degradation amongst the native masses, the potato ended in wrecking both exploited and exploiter.

★

Accounts of the route by which the potato came to Europe have been coloured by a remarkable number of mistaken assumptions and prejudices, considering the historically recent period in which the event took place. One of the most outrageous versions, pampering to school history-book mythology about Sir Walter Raleigh, has the gent only peripherally involved: From Virginia he is supposed to have sent a glowing description to his Queen and patron of a wonderful new food which could be prepared in as many ways as he could spell his name, and each a delight to the palate and digestion. Along with his praise went a stone of Virginia seed potatoes and planting instructions for the royal gardener. Up came the lush green plants with the bonny yellow and purple flowers, and the court was summoned to a banquet. The nobility ate boiled leaves and stem soup, petal pâté and butter-basted buds; the chef used every morsel save the tubers. Sir Walter Raleigh, as every schoolboy knows, was imprisoned in the Tower on his return from the new world, and wrote poetry on the eve of his execution.

This lovely story is full of howlers, not least of which is the fact that Sir Walter never set foot in Virginia and the potato was not cultivated in North America until it was taken there much later — by the Irish! But there are several stories of ignorant gardeners providing kitchens with the bitter berries of the potato plant: potatoes belong to the same family as Bittersweet and Deadly Nightshade and all their green parts are poisonous. The ignorance often extended to the method of propagation, for as late as 1765 members of a Sussex village, experienced only in growing vegetables from seed, fetched a man from another county on Lady Day (March 25th) especially to plant their tubers.

Some two thousand years after the first cultivation of the potato on the Peruvian coast, the Spaniards sacked the Incan civilization which then flourished. Along with vast quantities of gold, the spoils of conquest included potatoes, which the Conquistadores took on board as ship's stores. Tubers taken back to Spain were planted in private gardens and were described favourably as interesting new vegetables by various writers. Potatoes may have reached Ireland by accident, again courtesy of cook's stores on Spanish galleons. For when the Armada was dispersed, several ships were swept around Scotland by storm, and wrecked on the Irish coast, where they were plundered by peasants. But the influence of Sir Walter has not been definitely ruled out: Sir Francis Drake undoubtedly used potatoes as ship's stores during his various exploits in South American waters. When Drake left Cartagena in Colombia in 1586, he picked up Raleigh's agent Hariot and the other settlers whom Raleigh had sponsored, in Virginia. The timing of their return to England would have allowed Hariot to plant a few remaining tubers, and eventually to pass some on to Raleigh's estate in Ireland.

The potato arrived in Ireland at a propitious moment, at the close of what Salaman describes as the most fateful century in Irish history: "A century which had witnessed a hundred years of the tortuous rule of the

Tudors, culminating in the slaughter of more than half the population, the destruction of the homes of great and small alike, as well as
churches and monastic buildings. The countryside was ravaged, its
once famous wealth of woods and cattle squandered." In the face of
marauding bands of outlaws and soldiery who exercised a scorched
earth policy wherever they went, the potato was supremely attractive in that it was grown underground and could be stored underground. It was also easy to prepare and could feed both family and
livestock out of one cauldron above an open hearth of burning
turves.

For the Irish peasants who endured the campaigns of Cromwell
and William III, the potato was their only hope of survival, and in
times of relative peace they were in thrall to it through poverty.
Arthur Young describes a typical area on his tour of Ireland in 1776:
"Their food is potatoes and milk for ten months, and potatoes and
salt for the remaining two. A barrel of potatoes containing 280 lb.
will last a family of five persons a week." This would imply that an
adult man would eat around 12 lb. of potatoes a day. There were
years of failure, when severe frost or disease damaged the crop, and
the people suffered terribly. And then in 1845 came blight, which
devastated the crop two years running, and then again in 1848.
About a million people died from starvation and the attendant
diseases: dysentery, cholera and typhus. An enormous number of
tenants defaulted on their rents and were ruthlessly evicted. Within a
six-year period almost half a million people were turned roofless and
penniless on to the roads. A vast emigration to the United States
began and the population continued to fall until it eventually stabil-
ized at about 4 million, less than half of the pre-blight number.

Conditions in England and much of the rest of Europe did not
allow such a rapid and total acceptance of the potato by the labouring
class. Rather the new food remained for many years a welcome
novelty at richer tables, while labourers were suspicious and shunned
it. In 1619 the potato was banned in Burgundy on the grounds that it
caused leprosy, while in Switzerland it stood accused of causing
scrofula. In many places it was rejected because it was not mentioned
in the Bible, and was therefore deemed not designed by God as a
food for people; while during an election in 1765 there appeared the
slogan: 'No Potatoes, No Popery'. In 1774 the starving citizens of
Kolberg refused to eat from a wagonload of potatoes sent to relieve
them by Frederick the Great. And during the famine of 1795 at
Munich, great subterfuge was needed to trick the starving into eating
soup made with potatoes.

It was during the long isolation of Britain during the Napoleonic Wars that the price of wheat became greatly inflated and labourers were driven to eat potatoes instead of bread. Meat and cheese more than doubled in price as a result of the hastened enclosures of common land and loss of grazing rights, and milk became very scarce. But 1 acre of land under potatoes could yield as much food as 4 acres under wheat. This increased productivity could better support a growing population and at a time of food shortage was more profitable for the landlord. The upper classes exhorted the workers to eat potatoes. The clergy preached the potato's virtues; employers began the practice of allowing a 'potato ground' in part payment of wages; and a committee of the House of Commons clearly formulated a policy whereby workers should be weaned from wheaten foods proper to the upper classes on to a potato diet which would permanently reduce their standard of living. Thus the amount of potatoes grown not only increased considerably during the war, but continued to do so for decades afterwards. According to Salaman, given the high level of unemployment caused by the wave of enclosures, and the soaring price of bread, it was only the acceptance of the relatively cheap potato which prevented disaffection from becoming revolution.

The potato continued to provide a kind of dietary buffer on a national scale in times of high food prices and low wages and remained a mainstay of the rural labourer's diet well into this century. When grain again became scarce during the two World Wars the governments of the time waged highly successful compaigns to boost both the cultivation and consumption of potatoes. During the Second World War, the Ministry of Food guaranteed farmers a market for their entire crop at fixed prices, and encouraged them financially to draw on the latent fertility of their land by ploughing up pasture. The retail price was kept low enough, by Government subsidy, to persuade consumers to substitute potatoes for bread. There was also a strong appeal to patriotism in an intense propaganda campaign featuring 'Potato Pete' in all the media. Recipe suggestions involving potato chocolate cake and potato plus parsnip banana fritters may not have fired the public's imagination in the way intended. But an additional 500,000 acres were planted with potatoes, and the resulting harvests were utilized to maximum advantage.

When I called on Reg Dobbs of West Pinchbeck, just a few miles from Spalding, he blamed the potato campaign of the last war for causing one of the major problems on his land: "We try to keep our potatoes on a one-year-in-eight rotation if possible, because the trouble with potatoes in this area is that we grew so many during the war, when we were compelled to grow them, that we've got severe eelworm attack. It dates from those years. It doesn't affect the midlands and the southwest, and parts of Yorkshire, but it's widespread in the Lincolnshire fenland. The cyst remains viable for a very long time. So we're having to use expensive nematicides, and we have to rest our potato land as much as ever we can.

"Corn prices have been relatively stable for the last ten years, but potatoes are highly variable. We can go from very high prices one year to very low the next. There is no such thing as an average year with potatoes; every season is one on its own. There is an old saying in this area that 'pigs, peas and potatoes are a rich man's crop', because they are such a gamble. There is no certainty with any of them, either in the ease of production or the price you get at the end of the day."

Reg Dobbs described most of his neighbours as owner-occupiers who are financially sound, but he insisted that this is based on the profitability of grain production. Vegetables and potatoes mean a fluctuating income. He makes up his own rotation with winter wheat, sugar beet, peas, and bulbs and flowers, mainly daffodils.

The great variations in the price of potatoes from year to year stem from the variability in yield which in turn results from the crop's susceptibility to disease and climatic factors through the growing season. A shortfall one year may be followed by a glut the next. Difficulties in storing potatoes and transporting the bulky harvest, compared with grains, offer further complications. The Potato Marketing Board (PMB) was established in 1955 to help to manage this situation, and Reg Dobbs, elected by his fellow growers, is currently one of the 33 members of the Board. The PMB sets quality standards, sponsors research, and encourages exports and increased consumption at home. It manages a Government support price mechanism, whereby potatoes are taken off the market at a certain low level, denatured by the addition of a skin dye, and sold at a subsidized price for stock feed. But above all the PMB controls the acreage of potatoes planted.

Anyone who plants 1 acre of potatoes or more and sells their crop commercially must register with the Board. Reg Dobbs and his son Richard grow about 90 acres of potatoes on their 750 acres. "We have a quota and we have to pay a levy on it to the PMB which goes towards administering the scheme and contributes to buying up any surpluses. If we go over our quota we have to pay a levy at five times the standard rate, which discourages us from planting too many potatoes". A Devonshire farmer recently made quite a fuss in the press when he was caught by surprise growing a larger acreage than he had declared. The PMB inspector admitted that satellite pictures were commercially available, and implied that he had used them. "Unfortunately our quota is being reduced each year by about 5 per cent," said Reg Dobbs, "because the national average of potato yields is steadily increasing. It has now gone up to 35 or 36 tonnes per hectare, and a smaller acreage in potatoes can supply all the potatoes the nation needs. All over the continent potato consumption has been falling seriously but in Britain it is relatively steady. Our quota is about right for this particular farm. Around 10 per cent of total quota will change hands in one way or another each year, but a lot of farmers would like to expand. I'm very worried, too, about farmers whose dairy quota has been cut back. There just isn't a market for more potatoes. They can apply for quota and might be allocated some if they've got a good case but there just isn't much spare quota to go round at the moment."

Then I walked out to the Drainage Farm field where Richard Dobbs, in Wellington boots and waterproof jacket, was supervising the lifting and storage of Maris Piper. Rain at lifting time is a greater problem now that harvesting machinery is so heavy and can easily damage the soil structure. But the harvester is out on the field: rather the drizzle today than the deluge tomorrow. He ran through the crop's cycle for me, starting with the arrival of the seed potatoes.

"We buy in Scottish seed in November or December, and it goes straight into a new purpose-built store. We keep it heated against frost, and then we have to ventilate carefully in the spring. We get about 50 tons, in hundredweight hessian bags. Then we chit it in trays in the light, until the sprouts are about an inch long. Meanwhile we prepare the ground with a nitrogen phosphorus potassium fertilizer, and we put cow manure on about a fifth of the land to build up the soil structure. We store some bullocks for a neighbour, so we have a certain amount of muck. Then as soon as it's dry and warm enough in mid-April we go on the field with a two-row auto planter. It carries a lot of seed in a hopper and drops them at a pre-set distance according to the variety, and then covers them. We go over the rows again later to make up a nice ridge because we have to plant quickly when we can, and we don't always get it right. That helps to keep the weeds down, too, until the haulm grows full enough to smother them. We don't grow earlies because we're in a bit of a frost pocket.

"We spray with a systemic fungicide against blight early on, when it's taken into the plant to give good protection, and then later we use a leaf-type protector like tin compound. Then we irrigate most summers. They stop pumping from the drains in dry weather and that leaves us plenty of water — nearby. In September we spray with acid to kill the tops. We can't harvest until the skins are set and that takes about three weeks after the haulm has died. If we didn't kill them off early we'd never get off the land before the heavy rain, and we'd never get the winter wheat in. It means the crop is a little smaller, but it's dryer."

The Dobbs' land is spread over six miles and they have a mix of soils with some heavier clay, silt, and a few patches of peat. In terms of yields, potatoes actually prefer a heavier or 'stronger' soil, but this would aggravate the harvesting difficulties. "Probably we're growing too many potatoes for the land on this farm. We have to choose the siltier land but we're still lifting a lot of soil, and we're going on at a time of year when we should be getting off. In future we may exchange suitable fields with neighbours on lighter soils."

As I walked across the bare field my feet sank an inch or two into the soil at each step, and I expected my boots to double in weight within minutes, but surprisingly little stuck to me. And then I beheld the machine which caused the problems. The poet-philosopher John Stewart Collis, who was drafted on to the land in 1939 and found himself planting potatoes, soon saw machines in a new perspective:

I'm in no fit state to think it out — my back's aching too much. Empirically, as seen here regarding *this*, a machine seems excellent. And I fear that machines come into the world, not following a principle, nor with an eye to future developments, nor in relation to the whole, but by fits and starts, one by one, each seeming splendid to those concerned. I have to admit that whatever views I might hold in the study concerning mechanization, on this field, from this labouring angle, I would cast a highly favourable eye upon any man who appeared with a potato planter.

No doubt the potato lifter seemed equally splendid from the labouring angle, although in general machines tend to dispense with labourers as well as labour.

Ten workers were attendant upon this particular machine. One of them sat in the forward cab and was responsible for propelling the system down the rows at a suitable speed, which in these conditions was about two miles per hour. A cutting edge drove underneath the ridges in which the crop had grown, taking in two at a time, and all the material thus contained was propelled up an agitating conveyor system which shook loose the soil, small stones and 'tinies' which then fell through gaps back to earth. These small potatoes would be killed later by frost, or by the herbicides put on before the winter wheat. The larger matter was carried up into the belly of the beast where half a dozen women sat on either side of the belt under a polythene cover and took out clods, large stones and old sets — the original seed potatoes which had fed the growing plant and were now reduced to rotten carcasses. The ware potatoes rumbled on along a steep incline perpendicular to the course of the harvester and then dropped into an adjacent tractor-trailer, maintained in the correct position by another driver. A second tractor-trailer followed behind to take up duty when the first was full. And behind all of this, at a very leisurely pace, walked a youngster with a wicker basket. He reminded me of the man with the red flag who used to walk in front of the first cars on our roads; relegated, in these more enlightened times, to the rear. He carried nothing purpose-built, but the same wicker basket, in the crook of his arm, as the Kashubian peasant of a hundred years ago. His job was to collect the good tubers which managed to evade the system. He told me that he rode in the cab sometimes, when they were a person short. He liked that better.

Half-way across the field the procession came to a halt. The women lit up cigarettes and the three drivers collaborated in various operations on the undercarriage of the machine. "It's seized up," one of them told me. "Seized up with clods. That's what happens when it's wet." It took about five minutes to clear. "It's eight years old, this one. The boss has got two of them, and a one-row as well but we don't use that." He climbed back into his seat and started forward at once, shouting down to me: "The new machines they make now throw more sod out, but if they do foul up you can't get in to clear them . . . safety measures . . . too many people chewed up."

I followed the first tractor back to the store. There was a drum of rat poison by the door, emblazoned with the skull and crossbones. The barn was well-lined with straw bales, with circulation ducts built in at intervals for forced ventilation. The trailer tipped its load on the heap, which would be built up to twelve or fifteen feet later, with an adjustable elevator. Storage is now a sophisticated technical process. Proper ventilation and temperature control help to minimize bacterial rot and gangrene. "We use sprout suppressants," said Richard Dobbs, "and chemicals to heal wounds such as bruises, and control skin diseases." He said he was assured that these chemicals were quite safe after three weeks and then added, "They wash off, anyway."

There may be no such thing as an average year for the farmers who grow potatoes, but statisticians put a figure of about 6 million tons on our average British harvest. We import about 400,000 tons, the majority of these arriving from Cyprus, Egypt, Greece, Jersey, Spain and France in the period before our own early crops are ready. That gives us an average consumption of more than 2 cwt of potatoes per person per year. Eire is the only country with a greater consumption *per capita*, and only by a small margin.

Potatoes are grown almost everywhere in Britain but our principal commercial supplies of earlies come from Cornwall, Kent, Pembroke, Cheshire and Ayrshire, usually in that progression though following very rapidly on top of each other; and our primary areas for main crop are Lincolnshire, Yorkshire, East Anglia, and the Fife and Tayside district in Scotland. Most of our seed potatoes are also produced in Scotland, where a climate with strong sea winds and high rainfall is not amenable to the aphids which spread virus diseases.

Reg Dobbs explained the basic logic behind his choice of seed tubers: "We grow Maris Piper because they are in great demand; we have a trade for some reds so we grow Desirée; Romana because we believe that is good for pre-packing; Cara for the quality market; and Wilja or Estima for a second early, or an early main crop." But the factors involved in such decisions can be quite complicated: there is the variety's ability to "kittle", or produce many tubers, and to put on weight early, which will affect size and yield; relative resistance to diseases and drought conditions, and the ability to store well; the suitability to local conditions such as soil type, stony ground, climate, length of day; the eventual shape and colour of the crop; the ratio of water to dry matter in the tubers, which will govern their suitability for processing; and a host of qualitative characteristics such as texture and taste when cooked. Few other crops show such a wide variety of behaviour, and such ease of adaptability, as the potato.

The King Edward continues to command a premium price and is ideally suited to South Lincolnshire, though yields less than main crop averages, especially in dry years, and other disadvantages have led to a shrinking acreage. This potato was raised from unknown parents by a gardener in Northumberland who called it Fellside Hero. There was a great confusion of named types at the start of this century, with Up-to-Date, for example, masquerading under more than 200 different names. But the rechristened King Edward, launched commercially in 1910, was one of the last amateur-spotted 'chance' varieties, as systematic programmes began to concentrate on breeding for types resistant to disease. Donald McKelvie produced the Arran varieties and John Clarke, in County Antrim, the Ulsters. The Pentland prefix was used for varieties raised at the Edinburgh Plant

Breeding Station, Maris for those at Cambridge. The retailer is not obliged to display the name of the variety alongside the potatoes he has for sale, but this must be stamped, by law, on every 55-lb. bag. The county of origin is also stamped on, using a simple code. Most greengrocers are happy to look up this information for customers who want it.

I went to my father for one man's review of the most popular varieties. As a merchant who has bought and sold potatoes for over fifty years he is less prejudiced by a potato's performance in the field and perhaps more aware of its reception in the kitchen and at the table: "The business is always changing, and you lose some as well as win. Most of the old round potatoes with deep eyes, like Epicure, had a lovely taste to them, but they were too hard to peel. Majestics would do anything, they were a good, cheap all-rounder for a long time. They could look awful, with scab and all kinds of skin blemish, but under the peel they were always good. Still there were limits, even then. We used to get Arran Victory from Ireland in short years, which was purple. Unless there was a famine, it was like trying to sell dead men. It depended where you were, too. In the northeast they liked red potatoes like Kerr's Pink that fell in the water. But you couldn't give them away anywhere else because if you left them in the water a second too long they'd go to soup. In Scotland they still like a yellow-fleshed potato such as Record. But in England they've got to be white, like the new earlies, Ulster Prince and Ulster Sceptre, which are as white as driven snow.

"These days all potatoes have got to look nice in polythene bags, like Maris Piper. I think Piper is specially liked because it chips well. Fish-and-chip shops like it because it can stand in water without looking sad, it doesn't absorb so much fat, and it doesn't go soggy so quickly afterwards. Desirée is a good all-rounder, but it's a bit on the waxy side for me. In Lancashire they like a waxy potato, such as Duke of York in the old days. But I'm a floury man. A proper redskin planted on a red soil that it favours, like in the Eden Valley around Penrith, gave you a marvellous fluffy white potato. And of course, there's King Edward. This new Cara has a pretty splash of pink and eats well, but there's nothing can beat a King Edward off good Lincolnshire soil. The soil is as important as the variety. Scotch potatoes are usually good-looking, but they don't match for flavour. A good King Edward grown on a bit of muck doesn't need any meat with it. And it's a shame to dip it in fat."

Just as the taste of a potato is a product of the variety and the soil, fertilizer and climate in which it is grown, so is its nutritional value. But in general terms, potatoes contain about 76 per cent water, and 19 per cent carbohydrate, most of which is starch. This gives boiled potatoes an energy value of 23 calories per ounce, which has been interpreted favourably by double-headed advocates: as an attractive 'energy' food potatoes cost less per calorie than white bread, and yet as a high-fibre 'slimming' food they contain only a third as many calories as an equivalent weight of white bread.

Potatoes also contain about 2 per cent by weight of high-quality protein, providing 4 per cent of the requirement in an average diet, and assimilable quantities of numerous minerals, vitamins and trace elements. Citrus fruits have become our chief source of vitamin C, but potatoes make a significant contribution, especially early in the season. Scurvy was common among the population of British cities before their introduction. The eighteenth-century traveller who observed that potatoes were almost the only food of the Irish poor, and that it was not unusual to see ten or more naked children around the pot, may have made a simple logical error in claiming aphrodisiac powers for the tubers. But it is possible that a diet of potatoes and milk could provide all the nutrients essential to health.

Most of the nutrients lie close to the skin in potatoes, and for many years nutritionists have warned againt the wastage wrought by the peeling knife. The skins of old potatoes are perfectly edible. Vitamin C, in particular, is also diminished during cooking, gradually leaching into the water, along with other nutrients. The less time spent in water, and the sooner eaten once cooked, the better.

A good percentage of nutrients are preserved in most types of processed potatoes, where the problem is more likely to be one of addition than subtraction. Crisps, for example, have a much higher calorific value than boiled potatoes, because of the fat which they absorb. A very high salt content, which may be correlated with hypertensive diseases, has become associated with many kinds of processed potatoes. And the addition of artificial flavourings, colourings and other chemicals means that many potato products meet the general criteria of "junk foods". The days of "taytee and point", where you got a big enough plate of potatoes but the meat you just pointed at by way of indulging your imagination, are over. So may be the days of "spuds", derived from the special broad-pronged "spade" once used for lifting them. We have entered the age of french fries.

When I told Reg Dobbs that my father rated a Lincolnshire King Edward above any other potato, he seemed almost surprised: "Yes, I believe a Lincolnshire silt potato has a better texture than any other. I think there is a quality point here that we haven't made enough of." But it was when he himself raised the subject of processed potatoes that his eyes lit up and he began almost to wax lyrical. Crispers now use over half a million tons of potatoes each year, and more than three-quarters of a million tons enter the market as frozen chips and other parfried, chilled products. These processes require medium or large potatoes with around 20 per cent dry matter and low reducing sugar content, and research is increasingly focused on their needs. "This is the major change that is coming up. The processing industry is a major factor, taking 25 to 26 per cent of the total crop, and moving up. We've had a strong crisp industry for a long time in this country, but very recently there's been a tremendous expansion in the manufacture of frozen chips . . . french fries."

19

Roots — Carrots and Parsnips in West Norfolk

BILL KNIGHTS CAME to southwest Norfolk, on the edge of the Breckland, at about the same time as the carrot and the tank.

Abundant archaeological evidence suggests that in early Neolithic times there were more people living in the Breckland than in any other part of Britain. They were attracted, apparently, by the very light sandy soils which would have been easy to till with their primitive tools. But with a limited fertility, these soils would soon have been exhausted. Grazing animals such as wild deer would have kept the land open as the early farmers moved on and cleared another small patch of forest. The record at Hockham Mere, a lake on the edge of the Breckland which has gradually filled up since the last Ice Age with muds and silts and, finally, peat which have preserved the pollens of the ages, shows that in Anglo-Saxon times wheat, rye, flax and hemp were grown here, as well as a host of their associated weeds. With improvements to the tools of cultivation, and especially the introduction of the moldboard plough from eastern Europe in the seventh century, the light soil of the Breckland lost its special appeal as farmers took advantage of heavier, more fertile marls. By the time of the Domesday Book the area was only sparsely populated, and when bubonic plague reached England the severely reduced population concentrated in the most fertile areas so that on the marginal land of the Breckland twenty-eight villages were completely deserted. The Breckland is still sparsely populated today. The grasslands and acid heath are a delight for the botanist; they provide excellent territory for the military to conduct large-scale tank manoeuvres; and they have offered plenty of scope for the designs of the Forestry Commission with their massed serried ranks of pine trees.

Shortly after Bill Knights came to the village of Gooderstone, on the northern edge of the Breckland, however, the region also came into its own again with a specific agricultural advantage. It was discovered that the light, sandy soils with low organic content were ideal for the modern methods of cultivating the root crops, parsnip and carrot. Within only a decade or so the region had become one of the major sources of carrots for the entire country. The Knights family alone were growing 15 per cent of all the parsnips in the United Kingdom, eventually developing an export market for them and their carrots in Germany and Scandinavia.

"I was born in 1900 and in 1910 I started helping my father on our 7-acre market garden near Yarmouth. I was picking tomatoes, washing leeks, that sort of thing. The first year was good but then in the second we had a drought. My mother died and we had no money. So when I was thirteen I left school, and then when the war started our ploughman went off to fight and my Dad asked me to do it. I weighed four stone seven then. I was a little chap, but it was only a pony and I soon mastered the job. I was full of ambition, full of go for the work and everything.

"When I was fourteen, almost to the day, my Dad took ill and he said I'd have to go to market. I loaded all the stuff on the cart and off I went to Lowestoft and Yarmouth. Everyone was tickled to see me there and they'd climb up on the cart to see what I'd got, and they kind of looked after me. And then when my old Dad was better, he'd rather be in his greenhouse than going to market anyway, so I just kept on. By the time I was fifteen I was a market man and by the time I was sixteen I could go anywhere, do anything. One of the troubles with young lads today is they lack experience.

"We had good crops and cleared them, and it was lovely. Then after I'd met 'Mother' here, she went off to London so I started taking eggs up to London. I might have stayed but it seems I had working on the land in my blood, so I got going with chickens. We'd have 500 birds in one house and we'd live in the other. That was twelve feet by eight feet. That's how we started. I had no education. I never heard of decimals. But my father gave me a master's education in reading, and if you can read you can get to know anything. If you don't have all the education, you've got to use your head, and that's what I did on the chicken job: how to pick out the best layers, how to mix up the rations with all the right feedstock values. I had incubators in Norfolk before anyone else, so I got pullets laying in October when everyone else's had finished. It was a lovely job, the chickens.

"Then in 1931 farmers started going bankrupt and thought they'd try the chicken lark, and we ended up all going bankrupt. We owed £500 to the mill man. I told him I could sell up and pay him his money and have nothing left, but I said if he'd allow me I'd like to buy cows with what I had and pay him back as I could. A wonderful gentleman, he was. So that's how we got started with the milk job. We bought four cows. But there were 47 dairymen in Lynn at that time. No one would look at me and I got right downhearted and I went into a café where the chap said 'What's the matter with you?' and put me on to a fellow who'd pay 9 pence a gallon . . . at six o'clock in the morning and two o'clock in the afternoon. That was the first time he had ever signed a contract, and it took a week to get him to do it. Then by sheer luck I got into a retail round, pushing an old bike with sixty pints of milk on it. If you push a milk bike six or seven miles a day, as well as all your other work, seven days a week, three hundred and sixty-five days a year, then you know what work is.

"Then again by sheer good luck we got a farm at Leziate just before the Second World War, and then in 1942 we got this one here, with 500 acres. The boys didn't want dairy cows so we got sucklers. There were tanks running all over the place. I saw fifty-four tanks in one field once. They came right through where we'd set out the sprout plants one day, and right behind them came the Sappers with rolls of barbed wire, patching up the hedges again, but they couldn't patch up the sprouts.

"We've got 50 acres in the middle of the farm where there's a few feet of black peat on top of the chalk and when I saw twenty tanks at the edge of it I asked the chap in charge what he was up to. They were going straight across this field, he said, because 90 per cent of these tank crews were town people who always kept to the roads. That's what happened in France, he said, and they got popped off by the Jerries just like that, so they had to be trained to go through hedges and fields and anything, regardless. But couldn't he see it was real wet land, with all the rushes and such? I asked him. I reckon he must have been one of the 90 per cent, because in they went. They got twenty or thirty yards and then the tracks were going round and round and they were just sitting there.

"Then after the war my boys were round at the Young Farmers Club and someone there told Peter he knew of a man who wanted 50 acres of carrots. Peter came home and said: 'What about growing carrots?' And I said: 'Why not?' So we went in with this man from Wisbech and grew 50 acres for him for two or three years and when the land was finished we grew a few tons here. Well, we were digging carrots up with forks and there'd be bags left here and baskets there, a little patch left here . . . you couldn't work your damned land at all . . . so we got a bit fed up with it all. Then one Saturday the boys had got a great stack of canning carrots down the field and the chap

didn't come for them. So the boys had to take them. By the time
they'd taken 30 tons of carrots up to the cannery at Lynn the boys were
fed up with it proper. So Peter suggested we drop all this contract
work and grow them for ourselves and do it right. We grew 100 acres
that year, then 300, and we found it suited us, and it suited the land.
The canners wouldn't take big carrots, so we started selling for the
market. We started hiring more and more land and then we were very
lucky in being able to hire the whole of the Hilborough Estate.

"So that's how we went from 7 acres to 7,000.

"Work has been my hobby. When I was a market gardener, I loved
it. The chickens were just the same. Then the cows. And now it's
carrots and so on. It's marvellous. Fantastic. . . .

"I still get up between four and five every morning. I'll show you
what I do. . . ." And Bill Knights brought out the accounting books
for 'Knights of Norfolk'. There were rows and rows of immaculate
figures . . . columns brought up-to-date first thing each morning
. . . vast sums of money balanced to the penny . . . 25,000 tons of
carrots a year, all present and accounted for. "Marvellous," he said
again, not especially with pride but certainly with satisfaction.

Carrots and parsnips are both members of the Umbelliferae family
whose wild forms grow in grassy places in much of England and
Wales, although their tap roots are quite small, thin, and very pale-
coloured. Both plants were cultivated by the Romans. The first
parsnips were popular, but still very small, with fleshy types such as
we know today being developed only in the Middle Ages. The sweet
flavour of winter parsnips was much appreciated, and many sweet
puddings were made with parsnips as a base. They remained standard
issue with roast meat long after potatoes were introduced in
Elizabethan times and their green tops were fed to cows, for a
particularly rich milk, into the nineteenth century.

The first carrots were far less successful, and the Roman cultivars
appear to have been unwanted and lost. They were domesticated again
in a purple form, according to Edward Hyams, in Afghanistan around
AD 600. Seeds of this cultivar were gradually distributed through the
Islamic countries, and by the time they reached Moorish Spain a
yellow mutant had been selected and developed. Purple and yellow
carrots spread from there through Europe, reaching Britain in the
fourteenth century.

The yellow-type carrot went on to produce further mutants which
were selected and developed. One of these was a white carrot type
which was widely grown for cattle food, and another was the orange

type familiar to us today. The latter originated in Holland as recently as the seventeenth century. It was greatly preferred by cooks because unlike all the others it kept its colour when boiled, and looked most attractive. It quickly became established as a garden vegetable to the exclusion of all other carrot types, although some purple carrots are still grown in the Middle East.

Varieties of orange carrots continue to be specially selected for colour, so that now most commercially grown carrots are a uniform deep orange-red through their flesh and core. When the more hardy variety Autumn King was introduced from Holland in the 1950s for example, its considerable virtues were diminished by a rather too yellow appearance. Now our improved Autumn King types dazzle like the brightest Chantenay.

The colour of carrots is also one indication of their nutritional value. They contain moderate amounts of vitamins E, C and some of the B complex, and various minerals. And they are exceptionally high in carotene, which also gives the orange colour to apricots and winter squashes, and which the body converts to vitamin A. Amongst other vital roles, vitamin A is essential to the process by which our eyes perceive light. With a nationwide 'black out' for six years during the Second World War, carrots were much promoted as improvers of 'night vision'. Certainly the Ministry of Food was eager for any new angle which might increase consumption — the guaranteed price which they assured growers was fixed so high that the area under cultivation increased from 16,000 acres to 41,000 acres during the war.

This acreage subsequently dropped, but over the last decade consumption has increased by one-third so that carrots are now our third most popular vegetable, after potatoes and cabbage.

With mechanical improvements in root crop cultivation and the use of chemical herbicides and pesticides, carrots ceased to be primarily a market gardener's crop earlier this century, and were grown instead almost entirely by arable farmers as one of the root crops in their normal rotation. With the exception of small amounts of early carrots sold locally with their foliage in bunches, this is still the case today. Succeeding best on light friable soils which encourage germination, allow root penetration and easy swelling, and fall off the roots readily on lifting, carrots were considered an ideal crop for the Fenland soils of Cambridgeshire. They are also grown on the peat mosses of Lancashire, and on the siltier soils of the Trent Valley, including the Isle of Axholme, and what are known as the Humberhead Levels. But the business became dominated in the 1930s by a small number of large-scale grower-merchants in the Chatteris area of the Fens, led by

Arthur S. Rickwood who went from a 1-acre allotment at age sixteen
to 9,000 acres and was dubbed the 'King of Carrots' by George VI.
Fodder roots had been grown here since the coming of the railway, for
sale to the stablers of London and Midland horses, and as motor
transport came to predominate, carrot production for human consumption was stepped up. Farmers commonly hired out odd 'dirty'
fields for one season to root-crop growers, who returned them cleaner
and more fertile for drilling winter wheat. As these growers rapidly
expanded, they found the Fenland available to them limited and even
reduced, as the peat in some areas shrank with cultivation down to the
underlying clay which would not grow carrots. When they tried
renting and buying land around Norfolk's Breckland, they found that
the sandy soils produced higher-quality crops than did the shrinking
fen soils, and with locally based family concerns such as the Knights'
also expanding rapidly, this relatively small area now produces the
bulk of our national harvest.

The sandy soil has also greatly encouraged the other major
development in carrot-growing, which is the extension of the season
to cover virtually the entire year. With the introduction of the hardier
Autumn King types, carrots can be left in the ground, protected by
straw and black polythene, and harvested as required. This can only be
done on soils which drain quickly to allow mechanical access.

The last of the main crops are usually marketed in May, with
substantial pullings of early carrots at the beginning of July. The gap
is filled by imports from France, Holland, Cyprus, Italy and Texas,
and though these amount only to some 18,000 tons, against almost
500,000 tons marketed through the year, they attract very high prices.
British growers are now competing with these imports by growing
early carrots under plastic film covers. This technology has been
largely developed at the Arthur Rickwood Experimental Husbandry
Farm, under the auspices of ADAS, but it is not appropriate for peat
soils as they grow too many weeds, suffer too much frost-heave of the
surface, and give poor anchorage for the plastic sheets.

Most of these early crops, called 'second earlies' are planted early in
the new year, to be harvested in mid-June, but some 'first earlies' are
planted in October to over-winter as seedlings and then grow rapidly
to harvest at the very beginning of June. Six-foot wide sheets of plastic
costing some £350 per acre are 'floated' above four rows of carrots,
which are planted in depressed furrows so that the leaves do not touch
the film. The plastic, which is tucked into the earth along both edges
to secure it, increases the soil temperature slightly, and the moisture
from condensation can be controlled by a proper spacing of ventila-

tion holes. The film covers are removed in early May when the carrots have seven true leaves. F1 hybrids of the Nantes type such as Nandor, Nanco and Tarenco are best suited to this method. Nine growers, including the Knights, who represent between them around 90 per cent of early carrot production, have formed a marketing organization with the brand name Gold Prince. By regulating supplies and maintaining high quality they hope to bring what they call 'cohesion' to the market, and profitability to this fledgling crop, which is still expensive and risky to grow.

Bill Knights' grandson, Paul, took me from the packhouse complex out into a 45-acre field to see the main crop harvesters at work.

At eighteen, Paul is going soon for a working vacation near the Bay of Plenty in New Zealand, where his uncle Brian grows kiwi-fruit. In the meantime he is learning the business. A large percentage of the 250-strong workforce are employed in the packhouse, and most are women who live in the surrounding villages. They wash about 100 tons of carrots each day, as well as parsnips, for the daily deliveries to Tesco, and for the wholesale market. The high-pressure sprays used for washing the roots drive water into the fibres, according to some critics, damaging both taste and keeping quality. But Paul answered this, with more naïveté than cynicism, I thought: "Nobody worries about taste. It's the way they look that matters."

Out in the field, the carrots had little opportunity to show the way they looked. The only orange in evidence was the brightly painted metal of the machine as it slowly consumed the rows. In front of it lay a swath of frilly green leaf, and behind it, soft brown earth.

Not so long ago these carrots would have been dug up with a fork. Gang-masters organized bands of casual labourers, some of them experienced migrant workers who followed the harvests in turn. My father remembers giving explicit instructions to lorry drivers loading in Norfolk: pick up the carrots last, because his grower had hired gypsies. They were reputed to be excellent pullers, but they didn't start work until after teatime.

Paul's father, Chris, told me that a few growers still use gangs of women hand-lifting with forks, because it causes much less damage to the crop, but on his farm, and in this field, it was clearly the ethos of the machine. "We've got two top-lifters for earlier in the year when the land is drier, which pull the carrots up by the tops and then trim the leaf off before they're sent up the belt into the trailer. Then we cut all the tops off in the field as we go over with this machine, like a potato lifter, which digs underneath the roots and rolls them up a tumbling belt to lose the soil. They harvest a complete four-row bed. At three to five miles per hour they can dig 6 or 7 acres a day." Alongside the harvester, to collect the carrots, a tractor pulled a six-wheeled, forty-foot trailer with its own powered drive and immense wide tyres. "A lot of carrots break when they come off the belt and drop to the bottom of the trailer, but we've installed video cameras so that once the first off have made a pile the driver can steer his trailer so that no other carrot has to fall more than one foot. The first off we sell for stockfeed.

"These are a Chantenay type, drilled straight in the field in the spring. Then we spray with Vinamyl, which seals a crust on top of the soil to stop it blowing away. Hedges can help, but you've got to be very careful or your windbreak causes a huge draught and you channel it right down on to your soil. With seed at £120 per acre and fertilizer and herbicides, plus all the labour invested, you don't want everything blowing away in the wind. This product we use costs £60 an acre and we look on it as an insurance policy, just in case. It's a by-product of the paint industry and it breaks down after four to six weeks to leave a nitrogen residue, which boosts the fertilizer.

"We irrigate regularly from two-hundred-foot-deep bore holes, but in a dry summer yields can still be 50 per cent down. And we have to spray to keep on top of carrot fly. The EEC regulations are very strict, there mustn't be the slightest sign of carrot fly damage, so we end up doing it probably two or three times more than we would otherwise, just to be sure. DDT and aldrin were used in the past and people are worried about that now, but the new generation of chemicals are actually safer. There are no residues after about four weeks. They tested us once, when we sold some carrots for baby foods.

"In Holland they can fumigate the soil but that would be un-economical here, so we have to rotate. Otherwise we'd have problems with eelworm and a host of microscopic pests which accumulate in the soil. We grow 1,200 to 1,500 acres of carrots, and 800 acres of parsnips, and we rotate them with two years' pasture, two years of cereals and one of, say, turnips. This year the turnips got hit badly by cabbage root fly, which came over from the oil seed rape.

"The parsnips are harder to grow than carrots, and only yield half as well. They have large, flat seeds which tend to stick together and are hard to drill separately. And chemical weeding is more difficult. But there is a good, stable market for parsnips. We drill them from early March until the end of May, using a succession of varieties. There's Albino which is very white; Tender and True which is long and tender right to the core, even when it has grown slowly in a dry year; and several good new types developed in Australia, such as Melbourne. We pull some baby parsnips for cooking whole from the end of July and keep on with the main crop until May, so we've got ten-month cover. We use the same kind of machines as we lift carrots with, but we incorporate an air blower to separate off the clods and stones which come up with the longer roots."

These roots consist largely of starch throughout the summer, but within a few weeks of the first light frost the starch has mostly turned to sugar, greatly increasing their flavour and appeal. They are undamaged by even quite hard frosts and so are stored in the ground and lifted as required. The sandy soil suits them perfectly, for on wet soils with much organic matter they are prone to canker, caused by fungi reaching down from cracks in the root tops, turning black. And the considerable depth of soil allows the parsnip to become a kind of mineral-mine. The long tap root, which is trimmed off, often reaches between two and three feet deep, and through this the vegetable receives a good store of calcium, phosphorus, iron and other minerals. "They say parsnips are so high in iron," Chris Knights told me, "that when you cut one and leave it to go brown, you are watching a vegetable go rusty!"

20

Neither Muck nor Mystery — Brussels Sprouts in Bedfordshire

JOHN GUDGIN IS a market gardener *par excellence* and he talks of his work as of his "craft or sullen art". Flour was being ground on his 30 acres at Shefford Mills, Bedfordshire, in the thirteenth century and his own family has worked the land there since 1860. Through his London agent he has packed vegetables for Buckingham Palace and the Royal yacht. He has exported Cos lettuce to Germany and he consistently sends first-grade produce to markets as far away as Manchester and Belfast, as well as serving London and the Midlands. But following close behind this account of his achievements comes a voluble and diverse concern for the future:

"It used to be market gardening all the way from here to Cambridge, using every patch of soil for the right job. We've got heavy boulder clay over the canal, and clay again over the road. But on this strip it's a sandy soil with all the river bed deposits. It's top-class soil. We use Palmer's organic-based fertilizer, which has got fishmeal and so on in it, and we say so on all our printed labels, but it doesn't make a scrap of difference to the returns. You can't get manure now. We had 1,000 pigs at one time and all the muck went on the land. But pigs don't pay now. We only used to have to put a few hundredweight of potash on all the land, and now we have to use half a ton of compound per acre. The ground structure is gradually going. We get mineral deficiencies. And now irrigation is a problem. I've got pictures of my ancestors looking into the water and it's as clear as anything. But the stuff coming down the river has changed and now it's very murky indeed. The experts come and tell you it's okay but in dry weather it becomes more concentrated and when we sprayed the beans with water it burned them. My wife has a complaint called perforation of the bowel and they told us at the hospital that this is a common problem with farmers' wives from here to Peterborough, but they never said why. Why aren't they trying to find out?

"Eating habits have shifted. Nobody wants early peas now, and things like that. We've grown a lot of exotics over the last ten years . . . fennel, coriander, mooli, kohlrabi and things like that. Red cabbage, spinach and early vegetable plants. We grow almost everything. We have bees down every year to pollinate the pumpkins . . . 20 hives . . . so we're limited as to the types of sprays we can use. These big farmers grow corn and then speculate with a vegetable as a field crop. They can afford all the big machines. And big growers in co-operatives get subsidies. They can plant acres and acres of cauliflowers and then if the market is flooded they get paid to plough them in. The French get subsidies for their packaging materials . . . the Dutch get subsidies, and the Danes. The Germans get cheap labour from Turkey and Greece. . . .

"In 1950 we sold lettuce for 12s. a dozen, and a tractor cost £750. In 1985 lettuce makes just the same, 60 pence a dozen, and a tractor costs £10,000.

"We're down to 14 staff, and another 14 part-timers. Labour is a terrible problem. One man just died, and another's had a heart attack. All my men have worked here for over twenty-five years. There's no real open vegetable production left, for students to learn from. Almost all the market gardens round here have gone. There's no money in open vegetable production in this country any more. Last year was good, but the previous five were terrible.

"We can't grow our exotics to look as good as the continentals, but they taste better. They get less sun but are longer growing, and that shows in flavour. The plant breeders have given us the ability to increase yields but you can only work nature to a certain point. The weather is unpredictable. They're trying to grow vegetables like making a car. In the end they'll have to go back to traditional methods, and looking after the soil, but it won't come in my time.

"We've got the potential in this country for growing almost anything. And we've got the water, we've got everything. During the war we showed what we could do. We could be the seed bed of the world if we wanted to be. But all governments in Britain are industrial ones. The food industry, bar corn, lives on a knife edge in this industrial country and it looks as though it always will except in time of war. I've got three sons working with me now, but the future is very iffy. In twenty years vegetables will be like metals, then there might be something in it."

Six miles north of Shefford, at Sandy, Ken Quince has documented the decline of the Bedfordshire market gardeners in a booklet, *The*

Sandy I Knew. In 1960 Sandy was described in a gazetteer as having one *raison d'être*: 'market gardening'. Ken Quince and his son are still thus employed, but fifty-three similar family businesses have folded or switched to agriculture since the Second World War, and the Quinces have become almost solitary standard-bearers. For they take great pride in their work and status as market gardeners, and this exists independently of an undoubted nostalgia for a past which consisted of laborious grind and hard times as much as honoured skills and a love of the soil.

The one crop out of many on which the reputation of the Bedfordshire growers was especially built remains an important one in the area. Fifteen per cent of our national crop of 200,000 tons of brussels sprouts are grown in this small county, although the bulk of these are grown by 'farmers' as the fourth crop in a rotation with corn, sugar beet and potatoes.

The A1, the old London road, as it passed from Baldock through Biggleswade and Sandy to Duloe and St Neots, was once known in the trade as 'The Brussels Trail'. And as blown sprouts were heeled into the ground when pickers rejected them, to decay and put precious nitrogen back into the soil, it was said: 'and by its smell you shall know it'. But as I drove up The Brussels Trail there was only the smell of diesel oil and motor exhaust, from the heavy traffic on the road and the tractors and harvesters in the field. Today's F1 hybrids have been bred for the machine, so that all the small tight sprouts mature at the same time. The harvesters which deal with about three-quarters of the brussels sprouts crop drive through the fields only once. They cut the plant stem at the base and as it passes through the machine the sprouts are stripped off.

Just north of the Biggleswade roundabout on the London road in a field adjacent to a transport café, I spotted the three bent backs of Mr Perryman, Mr Peacock and Mr Smith. This was how they introduced themselves when I had woven my way through the waist-high plants, every leaf I brushed against tipping its reservoir of early October morning dew, or rainfall, on to my trousers. They smiled liberally at my sodden bottom half and agreed with each other that the wet was the worst thing about picking sprouts.

Messrs Perryman, Peacock and Smith work for the Quince family fulltime and are involved with the planting, cultivation and harvesting of dozens of different commodities. Picking brussels is the only job for which they are paid at piece-rate, and all three continued working as they answered my questions: "That chap ahead of us is taking off the tops. They eat lovely, brussels tops . . . just like spring cabbage. Then we come along picking the sprouts from the bottom as they're ready. It's the hardest job of all on your back, but it gets easier as you come through the field again later on, because you're picking higher up. We normally go through the same field four times. It can get cold by the end, but it's the rain that makes it hard. Wind and rain. We go on picking however hard it rains."

These three hand-pickers were quite dry beneath their all-encompassing waterproofs. They each carried a sheaf of bright green plastic nets tucked in their belts, and a large metal spring-scale. They picked straight into a net, which stood up in a plastic outer to keep it clean on the ground. When the net was full they hung it from their scale, topped it up to 20 lb., dropped in a 'class 2' label and tied the bag up. Then straight on to the next. They can pick a ton on a good day, and earn £30. "It's a decent wage," said Mr Perryman.

Ken Quince remembers the longer hours and small wages of his youth, when it was "understandable" that people were looking for more reward for their labour: "I remember cycling to Oakley to pick sprouts, twelve miles each way. You'd often have wet feet when picking brussels, as protective clothing was almost unknown, and if you wanted a hot cup of tea before going to work, the fire would have to be lit to boil the kettle, and keeping time for the working day was much stricter than it is today, and if you were unfortunate enough to lose time because of wet weather or severe frost, you were not paid.

"During the winter of 1929, after a very dry summer which almost killed the sprouts, 300 people were out of work in Sandy. Dr Campbell, a local G.P. and councillor, arranged work for some of them pulling sugar beet in Suffolk, staying on an airfield site, but I went into the fens digging potatoes. Agricultural labourers were not classified as skilled labourers and were therefore unable to draw unemployment pay. The local council found employment for the men in cleaning out the River Ivel. The whole work, which employed over 250 men, was done using hand shovels, with the mud from the middle of the river being passed along on planks until it reached the bank. Married men received £1 10s. a week, and single men were only allowed to work alternate weeks. As spring gradually approached, some work became available, mainly digging between rows of rhubarb. Wages were terrible, and I remember feeling very bitter at the time that the few bosses who could afford to employ labour certainly exploited it.

"I spent many days weeding carrots in this area. The seeds were broadcast-sown by hand — and would be considered ready for weeding when the first rough or true leaf appeared. A small hoe would be used and kneelers — pieces of sacking — would be tied around your knees. In the case of onions, particularly picklers, a bent knife was sometimes used. When carrot-bunching started in early June, we would go to work at four a.m. to pull the carrots, go home to breakfast at six a.m. and then wash and bunch the carrots for the various markets. These were fan-bunched and tied with string and part of the tops cut off. A really good workman could bunch 100 dozen in a day, and for the extra two hours' work we would be paid two shillings."

In his booklet, Ken Quince explains that most market gardeners had pockets of land in various places. The contrasts in soil texture from heavy clay to river deposits and very light sand was to spread the crops over the whole growing year, which was very important in terms of keeping a steady income. Every known vegetable was grown, although some individuals did specialize to a small degree. Hedley Marshall's land, for example, "usually only grew summer crops, white marrows for north country markets and runner-beans and perhaps just a few sprouts which Wilf Cope would gather and push to the end of Longford Road on his cycle. . . . Mr King near Western Way produced the Royal Sovereign strawberry for which he received a gold medal from the Royal Horticultural Society. . . . Near West Road it was first-class land which grew very early crops. . . . Then at Folly Farm crowds of women spent many weeks peeling onions. Many tons were peeled and put into brine barrels for all parts of the country. . . . In Bedford Road we had Mr Joe Cooper who concentrated on high-quality vegetables for Monro's of Covent Garden. . . . Acres of parsley were also grown in this area, the green top picked off in the autumn, sent to Glasgow to be used by the ton in the dye works there. When it began to grow again it would be bunched for the Easter markets. One firm of salesmen at Manchester with whom I dealt would sell 400 boxes a day."

As for the poorer pockets, such as the Heath in the Potton Road area, the crops would often fail, especially during a dry summer with no irrigation available. "Here it was known as 'Soldiers' Ruin' because many of the soldiers who returned from the First World War used their money to try and start a business. Almost without exception they failed, because of the poor nature of the soil; in fact it was described as being so hungry that it was dangerous to lay down your waistcoat, otherwise the soil would eat it."

Before the railways, Sandy's success as a market garden area depended on growers being able to cart their produce to London in a day, returning with a load of manure from the city's numerous stables. I had heard that the Bedfordshire sprout business, in particular, was based on the cut-rate deal which the London & North Eastern Railway offered for transporting horse muck out of London. Sixpence per ton was the cheapest rate for many years, and this would carry formidable quantities away from the infamous 'shit siding' at King's Cross just as far as Biggleswade and Sandy. Ken Quince remembers throwing manure from railway trucks on to carts using a special kind of fork, dumping it on the land at regular intervals and then spreading it again with a fork. To empty one truck of, say, 8 tons of manure was a very hard day's work. But in fact fresh manure was never used on sprouts. "We used train loads and train loads of muck, but you'd never sow muck for brussels, or it would send them maggoty.

"Worse than the manure was spreading soot, which was brought back from Manchester and Liverpool. This was sown on to the land where, because of its potash content and its dry nature, it was particularly beneficial to heavy soil. It was sown by hand and there only had to be a little wind and it would all blow into the next field. So it was often sown on moonlit nights, when there was less wind about. Eager as people were to earn a little extra money, there were only a very few who would take that on.

"Now all that good market gardening land has disappeared. The Sandy I grew up in was capable of growing the finest of vegetables . . . but it has been swallowed up, either with houses, schools or factories.

"We grow about 50 acres of brussels, but we're still market gardeners. It's not a matter of scale. A farmer grows three, or maybe four crops. We grow beetroot, spinach, lettuce, marrows, onions, cauli, broccoli, potatoes, stick beans, every kind of cabbage you can name and so on and so on. The brussels just fit in with everything else, with a very careful rotation because that's the only protection against clubroot, which is the ruin of Brassica. We sell locally but mainly our produce goes to the Bournemouth area, where we have our own business. It's 301 miles round trip. We've got our own wholesale depot at Christchurch and we serve shops and hotels down there. I don't know what we'd do if we had to send everything to market, with all the expenses."

In the past, Bedfordshire growers used to reckon that a pound of sprouts cost the same as a pint of beer.

"Of course, the whole way that people eat has changed. A woman isn't going to stand over the sink preparing vegetables today. Generally speaking a woman is at work and she wants a quick meal when she comes home. Only last Saturday we went into Marks and Spencers, and because it looks so clean and tidy, ready to go into the pot, it costs twice as much as it would on Bedford market. Sometimes I go into Sainsbury's and I'm quite disgusted at the prices the public are being charged. It's usually at least 300 per cent over the price we're getting. The public buys by appearance, or I should say the blessed shopkeepers do. And the public has to put up with what the blessed shopkeeper wants them to have."

In a more profound sense, appearance is everything for the brussels sprout. In common with many other visually distinctive vegetables, it belongs to the same species, *Brassica oleracea*, as the wild cabbage. From this thin and loose-leafed plant which is native to southern England and coastal lands around the Mediterranean, has been developed a host of different plants by selection to encourage certain characteristics, sometimes assisted by chance mutation. Selection for leaves with curling veins, for example, eventually produced the 'head' cabbages such as Drumhead, Savoy and January King; plants with a blue tint, indicating high anthocyanin content, eventually gave us the red cabbage; and stems with a swollen, pithy base, kohlrabi. Grown for two centuries in the Channel Isles where they double as a tourist attraction, the Walking Stick Cabbage can grow a fifteen-foot-long stem. Kale, collards, sprouting broccoli and calabrese are all descended from cabbage. Cauliflower was first noticed as a mutant of cabbage, segregated and propagated in Syria. It reached England in the seventeenth century via Cyprus, Italy and France, known for many years as Cyprus cabbage. The modern name derives from the Latin *caulis*, meaning cabbage. Similarly brussels sprout plants, with their dense, compact buds formed at the junction of each leaf and the tall stem, are thought to have occurred first as a mutant of Savoy cabbage. Mark Twain wrote that "cauliflower is nothing but cabbage with a college education."

Edward Hyams believes that this long process was begun in western Europe by the Celts, whose word *kab* is the origin of our word *cabbage*. Or that perhaps the invading Celts found wild cabbage already domesticated by one of the aboriginal tribes such as Ligurians or Iberians, who preceded them. The ancient civilizations of Europe placed credit elsewhere: according to Greek mythology King Lycurgus was caught destroying grapevines by the god Dionysos. While tied to a grape stalk prior to dismemberment, the king wept, and those tears which reached the ground grew cabbages. Roman writers, on the other hand, tell us that no food causes so much thirst as cabbages, because they grew first from the sweat of Jupiter which happened to fall to earth.

Whatever the heavenly or kingly influence, the cabbage has featured widely through Europe and the centuries in the diet of peasants. The Dutch gave us the raw 'coleslaw', the Germans, fermented, salted 'sauerkraut'. And across central Europe generations have survived through winter on black bread and cabbage soup. In the seventeenth century one writer described a typical meal of meat of various origins and quality served "with five or six heaps of cabbage".

These heaps may not, however, have been the over-boiled, tasteless and formless piles which haunt our memories of childhood and institutions. Dorothy Hartley insists that our modern 'conservative cooking' is as old as the hills and quotes this medieval recipe in support: "Take a large quantity of the warts — and shred them, and put butter thereto, and seethe them and serve forth — and let nothing else come nigh them." This kind of minimal treatment is essential with brussels sprouts to preserve their nutty flavour and a degree of crispness. It will also protect most of their high vitamin C content.

Not only were sprouts first noticed, selected and cultivated in the Low Countries, perhaps near Brussels, but they remained known only in that area for centuries. *Spruyten* were mentioned in the accounts of a Belgian market in 1213, and *sprocq* were served at a Belgian wedding in 1481, and yet these sprouts were virtually unknown in Britain until about 150 years ago. Now the United Kingdom is by far Europe's largest producer and consumer of brussels sprouts. Our summer rainfall is usually ample to ensure the healthy establishment of vigorous plants, while our winters are sufficiently mild to allow the development of the sprout buds through until spring. With this ideal climate for the crop we can achieve much higher yields than other countries. The co-operative group, Bedford-shire Growers, export ever-increasing quantities of brussels to Germany, Holland, France, Sweden, Austria and — to the delight of their PR people — the capital city of Belgium.

Bedfordshire Growers was formed in 1961 and with about 100 member farms includes many of the larger-scale growers. Its Managing Director, Bob Taylor, told me that nothing could beat the old 'Bedfordshire' varieties for flavour. These included Bedford Monach, Bedfordshire Prize, Bedford Fillbasket, and many others. They were typically large 'fluffy' sprouts, of a light green colour, with a very mild subtle flavour. The larger growers would save their own seed and guard the stocks jealously. These plants were slow-growing on the heavy soil and could withstand rough treatment and adverse conditions, holding up very well into the winter months. Also, being late maturing they did not clash with the harvesting of sugar beet and potatoes.

Bedfordshire sprouts still tend to come on the market later than those from districts with lighter soils, and build to their highest level in December. "A touch of frost does wonders for the flavour," is their slogan on the market. But now most farmers grow F1 hybrid seeds of Dutch origin. There are hybrid types to suit all the soils, from river gravel to heavy boulder clay. They produce small, tight-packed, dark green sprouts which are now preferred for freezing and by most other consumers. And they make mechanical harvesting possible.

These modern seeds also germinate more reliably, although sowing directly into the field in late April, which brings a large saving in labour, is still something of a gamble. Difficult weather in the first months can lead to gaps which need replanting. The drilling machine can also drop two seeds at the same station, which then need thinning. This 'mending up' of the crop by hand is very labour-intensive. So most farmers usually plant some seed under glass in February and then in outdoor seedbeds during March and April, setting the young plants out in the field in May and June. Problems then arise if the land is too dry at transplanting time, and irrigation is needed.

The standard spacing in the old days was three feet each way, but the smaller sprouts of the new hybrids flourish down to a spacing of two feet. Ken Quince grows his sprouts two feet-three by two feet, and says that everyone has their own 'perfect' spacing. Then a close watch must be kept to ensure that the crop is protected from its host of pests and predators. According to Bob Taylor: "There's probably no more difficult crop to handle in the field. But as soon as you mention 'sprays' you get the emotive response. We try to use varieties with a great degree of disease resistance, and we are learning a great deal about creating the optimum environment for the crop. If the plants are not under stress they are far less likely to suffer disease, just like humans. But we're still left with things like aphids. I don't believe the consumer is prepared to accept unsprayed produce — produce with aphid damage, for instance. There's always a three-week clearance period after spraying, before marketing, and once the cold weather comes there is less need to spray anyway."

Future improvements may well be made by breeding-in some of the qualities of the old Bedfordshire sprouts. "Most farmers are keeping a few acres of the old varieties, just in case," says Bob Taylor. "In the very severe winter of '81/'82, for instance, the traditionals performed far better than the new hybrids. So it's a matter of preserving the valuable gene stock."

21

From the Dung and the Dark — Mushrooms in Shropshire

FOR YEARS BEFORE the role of mycelia was understood, or spores identified, the market gardeners near Paris who first tried to cultivate mushrooms would cast old stalks and peel over the manure of their melon-beds, to be rewarded with a sudden bounteous flush of mushrooms a few weeks later or, just as likely, with nothing at all. If a casting-out of scraps coincided with a full moon, or a new moon; with dry weather or wet; or with any other condition which was conceived as possibly influential, and a good flush of mushrooms subsequently appeared, then these aspects of 'method' would be incorporated in the grower's personal secrets of success. Green fingers had little to do with it.

When I was a boy, my father aimed to sell 1,000 chips of mushrooms each week, and they arrived at his warehouse with an assured regularity. Many of them came by ferry from Ireland, or by passenger train in large metal cages from Selby in the Ouse valley. But the first time I went to collect them myself, as I scoured the tiny winding lanes of Staining, looking for the Mereside Mushroom Farm, I felt I must be stumbling along the trail of one of those mystery tours which operate from nearby Blackpool. I could see the tower in the distance, intermittently, through the tall hedgerows . . . I passed a working windmill and some beautiful country inns . . . and when I finally located one of the most successful mushroom farms in the northwest, all I could see was a cluster of low, windowless, cinder-block sheds and an enormous pile of manure.

When I asked the manager, "Why here?" he answered that mushrooms like it "warm, wet and windy". That is probably a far less mysterious description of the local climate than any of the fantasies issued by the promoters of Blackpool as a resort.

Now I chose to visit the mushroom farm closest to my home, which is just a few miles down the road, near Hinstock in north

Shropshire. Our local climate can certainly fill two-thirds of the preferred "wet, warm and windy" conditions, but I soon learned that the growing environment is thoroughly and minutely controlled, whatever the state outside. Rather than green fingers, the owner and director of Kingcup Mushrooms Ltd, Peter Munns, has training in the exercise of engineering precision and scientific control and management.

"I've got no background in farming at all, my background is in engineering and chemistry. I took a position in 1965 as a chemist and research aide with one of the larger companies who wanted to set up a control lab and carry out routine analysis. I was soon put in charge of all growing operations. I was able to combine my previous experience as a chemist in brewing and malting, and in iron and steel, with the various process-engineering skills which I had acquired during my time in those two industries. After two years I became general manager for quite a big company and then I became managing director of a group which was the fifth largest production unit in the country. You've got to be able to understand engineering principles and have a fair appreciation of chemistry, biochemistry, biology and bacteriology. You've got to be able to put together all of the factors that apply because it's generally a far more scientifically-based type of growing than most.

"In 1976 I purchased this farm off Mr Knowles, who had left the colonial service in 1963 and come here with his little pot of gold and set up a mushroom farm because he'd always wanted to grow mushrooms. In those days the farm produced 1,250 baskets, or just under 4,000 lb. of mushrooms per week, and I've been able to build it up to over 20,000 lb. It would be difficult to set up today with a small amount of capital. In Ireland there is extensive, small farm production, but in England mushrooms are grown intensively and the larger companies control the distribution of the majority of mushrooms. There is scope for small firms but they're always subject to lower price returns because they'll only have the open market available to them, and mushroom marketing is moving away from that. Only 20 per cent of my mushrooms go on the wholesale market, for instance. The rest go to multiples and direct sales. We deliver them to Tesco's depot at Worcester and from there they go to south Wales and the southwest.

"You can grow mushrooms anywhere, if you can control the environment, but I think that anyone who grows at sea level has a slight edge on anyone who grows at a higher elevation. They have a more constant humidity and the effect of the sea bringing warm inshore breezes at night to give a more constant temperature. So they have a slight advantage with the ambient temperature and conditions."

Various types of fungi were no doubt gathered, eaten and enjoyed by the earliest of our ancestors. The deadly poisonous character of some, and the hallucinogenic effects of others, were also no doubt experienced from the earliest of times. Certainly mushrooms feature prominently in the folklore of many peoples. There are ancient fables constructed around an argument as to whether mushrooms are animals or plants. Their mysterious and sudden appearance (one species can emerge and expand eight inches in one and a half hours) followed by a similarly rapid decay, has inspired local legends of Satanic significance. Equally, they have been seen as symbols of the bounty of Mother Earth and the embodiment of fertility. Ancient Greeks and Romans considered them one of the foods of the gods, though some of their writers thought otherwise. Understandably, perhaps, as he lost two sons and a daughter to mushroom-poisoning, Euripedes warned of their dangerous character. Seneca wrote that mushrooms "are not really food, but are relished to bully the sated stomach into further eating". And yet some varieties obviously have served as real food. Darwin, in *The Voyage of the Beagle*, observed: "In Tierra del Fuego the fungus in its tough and native state is collected in large quantities by the women and children and is eaten uncooked. . . . With the exception of a few berries . . . the natives eat no vegetable food beside this fungus."

American writer Waverley Root claims that the Japanese have cultivated a fungus called *shii-take* for at least two thousand years, but the first record of mushroom cultivation in Europe dates from the 1600s when Oliver de Serres, agronomist to Louis XIV, began experiments with the field mushroom *Agaricus campestris*. In 1678 another Frenchman demonstrated that if the white threads which develop in the soil under mushrooms are transplanted into good compost, more mushrooms will subsequently appear. For many years this 'flake spawn' was dug up from the sites of wild mushrooms and transferred to waiting beds. 'Spotters' of particularly rich, strong strains would rush it by train to the nursery where it would be used to inoculate special bricks of manure. This spawn was not always reliable, and would often carry diseases within it, but one French grower in 1868 was able to produce 3,000 lb. of mushrooms daily, from twenty miles of ridge beds in caves near Auvers. Arthur Linfield began growing mushrooms in 1884, using the space beneath the grapevines in his greenhouses on Chesswood Road, Worthing.

In 1893 spores from the gills of mushrooms were finally identified, with the aid of the microscope, at the Pasteur Institute in Paris, and from these a pure-culture spawn could be grown on a nutrient jelly in a laboratory, later to be impregnated on sterilized rye or millet grains to be sold to the growers. The French kept the process for manufacturing this pure spawn secret. The United States Department of Agriculture developed the technique independently in 1900 but it was not until 1932 that pure spawn was finally manufactured commercially in Britain. From that date, with reliable spawn and a tariff on imported mushrooms, cultivation on a large scale could begin, although it was not until after the Second World War that the commercial companies themselves 'mushroomed', to the point where they now produce 200 million lb. per year, or almost 4 lb. per person in the country.

In fact, the vast number of spores produced by mushrooms enables them to be seen by the naked eye via a simple experiment which any child can conduct, alongside his saucer of mustard and cress on blotting paper or his avocado stone suspended with pins from the top of a jar of water, offering a dramatic progression in scale. An ordinary edible mushroom produces an average 16,000 million spores, or 'seeds', which are successively 'shot' off the sides of the gills by a carefully calculated 'explosion' at a rate of about 100 million per hour. If the cap of an open mushroom is cut off its stem and placed gills-down on a piece of white paper, with an inverted glass jar placed over it to protect it from draught, within about 48 hours a perfect print of the gills is produced by the falling brown spores.

In the field these spores fall among grass and are perhaps eaten by horses to pass out again in droppings which provide a suitable medium for germination and growth to take place. This growth consists of the mycelium threads, or spawn, which gradually branch out and digest their way through the medium. The spawn uses up most of its food supply but stores some in reserve. Under very precise conditions of temperature and humidity the mycelium organizes this reserve into tiny fruiting bodies which suddenly emerge and 'mushroom up' to produce more spores, and continue the cycle. Mycelia may live for a few months or hundreds of years, depending on the food supply, which explains why you can return to the same place each year with a good chance of success. In the wild, fruiting usually occurs only once a year. In the case of the 'field' and 'horse' mushroom, this is typically during a spell of warm, wet weather in early autumn. The mushroom used in cultivation, *Agaricus bisporus*, is almost identical to the field mushroom, except that its spores are produced in pairs instead of fours.

Peter Munns started his tour of Kingcup Mushrooms in the large concreted yard where the compost is initially prepared. There was a tall, fifty-foot-long wall of baled straw and at the far end a large, wet pile seeping a brown fluid which had a rich though not obnoxious smell. Next to this was another, sweeter pile at a more advanced state which was being fed through a large machine which aerated and wetted it. A very neat block some seven feet high, five feet wide and at this point twelve feet long gradually emerged at the far side of the machine and a man with a light, two-pronged fork walked around constantly, patting the outer edge into shape with the attention and finesse of a sculptor.

"We buy in the straw and wet it and spray chicken manure from a slurry pit over it. Then it's mixed and turned and topped off with horse manure which we buy from local stables and from a manure merchant who ships it in from up north. It doesn't smell because it's an aerobic process, with lots of aeration. Every two or three days the heaps are re-formed so that we can adjust the water content and replenish the oxygen, to enable the bacteria which break it down to function. We keep turning the outside into the middle and so on, to get as even a composting as we can. It will heat up to as high as 180°F in the centre, so that in thirteen days it's turned from straw to compost piles. Then various nutrients are added, such as nitrogen (in the form of cottonseed meal) and gypsum, and are turned into the stack. Then the compost is loaded into the trays and taken inside.

"Mushrooms always used to be grown on beds about two feet six inches high and three feet six inches wide on a concrete floor. But they had very little temperature control over the mass of compost and they were subject to infestation by pests from the floor, so then growers started using a system of fixed shelves. You could get a lot more compost in if you had shelves eight deep, but all the filling and emptying was done manually and it was quite difficult. After the Second World War English growers started using Grimsby fish trays and old ammunition boxes which they would stack in a chequerboard fashion, edge on edge. Then the Americans devised a large box on legs which would stack up, which brought us into a situation where we could move a great mass of compost using fork-lift trucks."

These heavy wooden trays on foot-high legs are filled to a depth of about nine inches with the compost and stacked four deep in a shed for pasteurization. Fresh air is blown over the trays for three days, cooling the compost to about 130°F and supplying oxygen. Peter Munns grabbed a handful of compost from one of the trays and crumbled it to reveal white thermophyllic bacteria amidst the fragments of straw and the black friable matter to which much of it had been reduced. "This process selects out the organisms which thrive, and they pasteurize the compost. It's the residue from these bacteria which actually provides most of the food for the crop in the early stages."

Once the compost is pasteurized, the mushroom spawn is added. At Kingcup they buy-in hybrid spawn from France and Italy, although there is a British firm which grows spawn. He showed me plastic bags containing 5 lb. of rye and millet respectively, each thoroughly impregnated and suffused with thin white threads of mycelium. "The hybrid is a blend of the white and the cream strains, giving us the advantages of both — the marketable whiteness of the one and the tissue density and ability to stay closed longer of the other."

For two weeks the mycelium grows through the compost, and then the trays are covered with a two-inch layer of peat and gypsum which is called 'the casing'. Each 'shed' in the complex of buildings is insulated and can be sealed off to form a mini-environment but all are accessible for mechanical handling and movement of the various media. We dipped the soles of our shoes in a sterilizing fluid as we entered the complex and from that point on were alert to the rapid movement of fork-lift trucks along and across our path at many points. Casing material fed through a large hopper in one wide bay was carried along a conveyor-belt on a formidable machine which deposited an even layer on to trays of waiting compost, which were then speedily returned to their sheds.

"Mushrooms give off quinones, which prevent a crop from forming on top of its own mycelium, so one of the purposes of the casing soil is to mop up quinones. There are no nutrients in it and therefore no reason for the mycelium to spread up into it. To encourage the growth we run the beds at about 80°F with a humidity of 100 per cent and a high carbon dioxide atmosphere. It's possible to computerize these controls with various degrees of sophistication, and although the high humidity can cause problems with currently available sensors, this technology will obviously be developed.

"After about three weeks' growth through the casing, the mycelium will start to send up its fruiting bodies, which appear as a mass of tiny white pinheads on the surface of the beds. They double in size every twenty-four hours and in five days from pinning the first flush is ready for picking. We'll pick these over three or four days then the bed will lie bare for three days until the second flush appears. We get 50 to 60 per cent of the crop from these first two flushes. We pick three more flushes after that, but it takes progressively longer. The main thing is to get the crop off as quickly as we can. In nine weeks the cycle is over. We crop 5 lb. of mushrooms per square foot in that nine-week period. Then the compost is watered over with an algicide and sold to gardeners — £39 per lorry load, and it's booked up a month in advance."

There are 30 permanent staff, full and part-time, and 50 pickers. Most of these are married women, many of them from the army base nearby, who work some twelve to fifteen hours each week. They hook a small shelf on to the side of the tray they are picking for the four baskets into which they must grade the crop: closed cups of different sizes, opens or 'flats', and seconds. They pick the mushroom, cut the stalk neatly with a knife, and select the grade as they go along, climbing on to the edge of the bottom tray to reach the top one, which stands eight feet from the ground. When I spoke to a picker later, she told me that it was not an unpleasant job. She knew of some families where they put the woman straight in the bath when she came home, but although the sheds are filled with a rich aroma of mushrooms she never noticed the smell once she left. "It was odd sometimes to spend all day in the warmth and dark and then come out into the world and not know whether it had been sunny or raining or whatever. And it was harder once you were out of the three months' training and on piece-rates. You had to pick 27 lb. in an hour to earn £3.50 and it didn't have anything to do with nimble fingers, but how good the beds were. There would be poor picking some days, on the later flushes, or when you were just cleaning up the beds. Then when the mushrooms were there, they'd have to be picked so you'd get a phone call at eight in the morning asking if you could work that day, and you'd be expected to."

Peter Munns says that labour costs represent 48–50 per cent of the total cost of production. He also says that good profits are hard to come by. "The Dutch have developed the older shelf system using modern techniques and materials and made a very fine growing system. But they've taken it from a system that was moderately labour intensive to one which is very heavily mechanized and capital-intensive, and the price has to be paid. Another possible way out is the deep trough system developed quite successfully by John Lockwood, a small British grower at Ripon. This system is a lot less capital-intensive and cuts down on labour. It consists of one three-foot deep trough of compost and you create a mini-environment, blowing air through it, so it's far cheaper. Also you do everything in a single-zone situation — peak heating, spawn-running, casing and everything — and then just go in with a scoop or push it out when you're finished. . . . Then there is the possibility of mechanical harvesting, but those machines cause a 15 per cent crop loss. They are used mainly where the crop is going for processing. The French and Germans eat far more mushrooms preserved than fresh, but in Britain it's the other way round. The trouble is that as the mushroom stem grows longer, it breaks the veil and the mushroom opens. If the geneticists can work on this and give us strains which stay closed even when the stem grows long, then mechanical harvesting will come."

Another way to increase profitability is to increase demand, and the Mushroom Growers Association (MGA), to which Kingcup belongs, wages numerous promotions with this purpose. Formed in 1945 and now a specialist branch of the National Farmers Union, the MGA has about 1,000 members, in 55 other countries as well as Britain. In one year recently over half a million pounds was spent advertising the single message: 'Make room for the mushrooms.'

Increasingly mushroom promotions boast of the product's nutritive value. In the past, writers have usually dismissed this aspect of the mushroom, concentrating entirely on the nuances of flavour and texture it confers as a food adjunct. The 1970 edition of the *Encyclopaedia Britannica*, for example, states that: "Nutritionally, mushrooms are vegetables of insignificant food value." It proceeds to list the constituent parts as more than 90 per cent water, less than 3 per cent protein, less than 5 per cent carbohydrate, less than 1 per cent fat and about 1 per cent mineral salts and vitamins. These same figures are turned to advantage in the modern promotional literature. Mushrooms are "high in fibre and low in fat, with no cholesterol". They are "low salt" and "low calorie with no sugar or starch, so they justify their place in any slimming diet". At the same time "they contain more protein than almost any vegetable" and "they are richer than most vegetables in vitamins B1, B2, and B6". These nutrients are provided in the average diet, of course, by foods outside the vegetable group, but no doubt mushrooms can contribute usefully to the total vitamin and mineral intake.

Since we are encouraged to view mushrooms as a health food, and especially as we are advised not to peel or even wash them — simply wiping them over with a damp cloth — I asked Peter Munns if he could assure consumers that they were free of any chemical contamination resulting from their cultivation.

"As far as my farm is concerned they are reasonably organically grown. We use no fungicides and the only things we use very, very occasionally are smokes between flushes to control flies. We don't apply anything to the beds at all. We clean out the sheds with high-pressure warm water and then with chlorine — the same stuff that is used for sinks and treating town water supplies."

Then I asked about the other recent trend in consumer interest, which is the desire to try out a range of varieties of different commodities. The cultivated mushroom is unique in its singularity. More than 5,000 different species of fungi exist in Britain, of which at least 1,000 are edible, and yet our growers cultivate only *Agaricus bisporus*. The situation is the same in the USA and even in China, which is a leading mushroom producer with a staggering output of 100,000 tons per year, most of which are exported. And yet about 80 species are eaten in France, though only a dozen or so of these appear on the market. And just as flat mushrooms are more flavourful than the more widely available cups or buttons, so those who collect a wider variety from the wild are usually delighted by the richer taste they enjoy.

"Yes, inevitably we will have more variety. In fact one larger grower has already tried this, but it hasn't taken off very well yet. We are looking at the Oyster mushroom and the Paddy-straw — the Volvaria — but we haven't started work with them yet. They will be quite different in appearance, and somewhat in flavour. You see, the French and Germans and Italians have the advantage of many more varieties of wild fungus which grow in greater proliferation than they do in England, and they tend to harvest them from the woods and process them and sell them. We don't have the large tracts of woodland and the sorts of summer and autumn temperatures that support the development of these. British people are more cautious of eating wild fungi."

Not all fungi will prove amenable to cultivation. The truffle, which grows underground in beech and oak woods and is 'sniffed out' by pigs, has defied centuries of human effort at exploitation. One sample changed hands at Covent Garden recently for £85 per pound, and I am sure that is not a record. Perhaps the orange Chanterelle, the pale-brown Morel with its deeply mottled ridges, the lilac-tinged Blewit, the spongy yellow-green Cep, the Shaggy Parasol with its pretty brown scales, and even the Giant Puff Ball will be more amenable. But they are likely to remain exotic lines, either free from the wild or highly priced on the specialist market. The Oyster mushroom, with a very short stem, yellow gills and a blue-grey cap shaped rather like an oyster shell is already cultivated on the continent and is likely to be the first widely available alternative mushroom in Britain, once the supermarkets decide to promote it and stimulate demand.

A more likely alternative crop for mushroom growers in the future is not a fungus at all but the increasingly popular salad vegetable, chicory. This idea is not attractive to Peter Munns, as the time when it is most seasonal, early in the year, is precisely the time when mushroom sales are at their highest, although he thinks it may be a sensible use for older sheds no longer suitable for mushrooms. Chicory is another commodity with something of a mysterious and secretive past. But the firm of A. G. Linfield now at Pulborough in West Sussex, who market mushrooms under the 'Chesswood' label, have developed chicory cultivation to a high degree and boasted until recently that at one point they were actually the largest single producers in Europe.

The loose, green, bitter leaves of the wild chicory plant were eaten by the early Egyptians, Greeks and Romans. Cultivated by the latter to obtain a less bitter salad vegetable, this is the green, shock-wig salad we call endive. The carrot-like root of the plant was used medicinally, and from about 1775 was roasted and powdered to produce a coffee-like drink. Chicory roots are still grown in the fens to provide a coffee 'stretcher'.

In 1850 a man called Bresiers who was head gardener of the Brussels Botanical Garden tried to force chicory plants, transplanting some root cuttings into his mushroom beds. Instead of the usual loose heads he obtained tight, white-leaved shoots resembling the hearts of cos lettuces which are now called 'chicory' in Britain. In the United States this is called escarole. In Belgium it is *Witloof* (meaning white leaf) and in France, *endives Belges*. For the crop was for many years a singularly Belgian affair. Monsieur Bresiers experimented to perfect his discovery and kept his methods secret for the rest of his life. Eventually his widow told her gardener how to produce the shoots and the technique gradually spread to neighbouring families. The vegetable first appeared at an exhibition in 1873 and was finally offered for sale at auctions only in the early years of this century. A crop grown by small family businesses in the province of Brabant around Brussels, where refined techniques were guarded jealously and passed on only from father to son, it came to dominate Belgian horticulture. It is still their most important single product, with an annual production of some 80,000 tons.

Seed is typically sown in the open in May or June. The plants are lifted during October and November and replanted in specially prepared soil under heated silos made of corrugated iron, looking like miniature

Nissen huts. The leaves are cut off an inch above the root and the crowns are covered with light soil or peat and straw. The new sprouts are forced in a blanched form through this covering medium, which must be carefully removed before cutting. Two or three crops can be cut from the same roots.

At Chesswood, according to their sales and marketing executive R. L. Haycock, they developed the hydroponic production of chicory and over a five-year period achieved considerable success. The roots were lifted in the autumn, and the majority were stored at temperatures close to freezing, using the facilities of a nearby apple cold store. They were then brought out as required and stacked together, much like the early rhubarb roots at Wakefield, in watertight trays in the mushroom sheds. With all the nutrients required for the growth of the shoots stored in the roots they were simply warmed in water in the dark, although the equipment and conversion of the sheds for this process were quite expensive. By this means Chesswood became the first chicory producer in the world able to produce commercial quantities all the year round. With a continuous demand from several supermarket chains they were poised to recoup the costs of development and become highly competitive when their parent company axed the programme.

The big food companies are themselves pioneering another aspect of fungus cultivation, although it can hardly be described as mushroom-connected, and Chesswood is not involved at all. This is the less-futuristic-than-we-think project whereby fungi at the mycelium stage are cultured in a liquid medium to provide protein-rich bases which can then be spun into various food substitutes, rather as is already done with protein from soya beans. With the addition of appropriate flavourings and other chemicals we can then have fungus sausages, fungus burgers and the like, although no doubt the sales people will devise more acceptable names.

"This is all very developmental and hush hush," I was told. "It's an industrial process and it has nothing to do with farmers or growers in the normal sense. It's protein and vegetable production for the twenty-first century. We won't be producing carrots, for instance, but carrot cells in H_2O . . . to be formed into a carrot shape at some advanced stage. . . ."

22

In the High Street

A VEGETABLE GROWER's Utopia might consist of 100 acres in the Vale of Evesham with half the land Lincolnshire silt and the other half fenland peat, none of which had ever seen a brassica, an onion or a potato. But farming and horticulture, perhaps more than any other profession, is the art of the possible. The farmer works traditionally with the earth — with the soil on its surface, the weather which wraps it, and the myriad of competing organisms which also inhabit it — and the earth can be both a stern and fickle medium. Some growers believe that they have taken charge of the equation by eliminating variables — including, in some cases, the soil itself. But however ideal or 'controlled' the conditions, the grower lives with an apprehension of failure until he brings his harvest home. And even then he can meet with ruin. Jonathan Swift wrote, "Whoever could make two ears of corn, or two blades of grass to grow up on a spot of ground where only one grew before, would deserve better of mankind, and do more essential service to his country than the whole race of politicians put together." But this offers little comfort to a grower who must plough his produce in, because the market is over-supplied.

For in general our markets *are* very well supplied. The host of factors which govern conditions in the great fruit and vegetable districts — the type of soil and underlying rock; the degrees of soil moisture, sunshine, frost, and wind; the elevation and aspect of the land; the inheritance for good or ill of previous cultivation; the nature of land ownership or tenure; the traditions and expertise of current management — all these lead, in general, to an abundance of produce in the high street.

I stood before the heaped pyramids of apples and pears, carrots and parsnips, tomatoes and cucumbers, in a small greengrocer's shop in the town near my home. In northern Shropshire, close to the borders of Cheshire and Staffordshire, Market Drayton has a population of about 8,000. There is a little light industry in the town; there is a percentage of professional workers who drive about twenty miles to work in the four surrounding larger towns or cities; there is a level of unemployment higher than the national average. Above all, Market Drayton is a market town serving the rural population for a large area around it. Mike Sutton is twenty-six years old and has been running 'The Country Kitchen' at the centre of town for three years. I visited his shop towards the end of the year when foreign produce was beginning to predominate, with the first displays of Christmas nuts and dates, oranges and satsumas. Even so, there was an excellent range of British-grown produce.

"It's apples and pears mainly, as far as the fruit goes. We're offering Cox and Egremont Russet in two sizes, Spartan for a red, Laxton Superb, and Ellison's Orange. We haven't handled Crispin up to now because although there's a good market for it we have to limit our lines. Unfortunately we just don't have space for everything. For pears, we've got Conference and Comice. We do English Beurre Hardy and Williams occasionally, although the latter especially has a very short season.

"We've got swedes and white turnips; a dirty carrot from a local grower and a washed sample, together with parsnips, from Norfolk; we're getting dirty celery at the moment from the fenland near Ely, you can tell from the black, peaty soil on it; brussels, leeks, spring cabbage and Primo, red cabbage and a lovely sample of calabrese from the Vale of Evesham; there's hothouse lettuce with the lighter coloured leaf from Evesham, too — I have a tendency to go for produce from the Vale because I used to live there and my wife's family comes from Chipping Campden, so I feel that I've got connections with that part of the world. The Savoys are from a very good growing area around Lichfield, and the caulis come from Lincolnshire. A lot of caulis have been lost just recently, but ours are still frost-free. The spring onions are starting to get a bit tatty — you can see the tops starting to go. They don't like the damp weather. You've got a choice at this time of the year — you either have tatty spring onions or you don't have any at all. The bulb onions are from Lincolnshire, with the Spanish on offer for people who want a mild onion for salads and so on. And we also run English picklers. The garlic is actually Spanish because whenever I see English garlic it's pre-packed in threes and my customers only want to buy one bulb at a time. English 'cumbers, Capsicum and tomatoes finished just last week, and courgettes two or three weeks ago. We'll start up with them again next March or April and run them all through the summer. Mushrooms come from all over the country, sometimes from a local grower. I choose whatever quality looks best on the market. We shall have ice-packed watercress tomorrow morning. We normally have about 60 different lines of produce in the shop, and in summer up to 50 or so will be British-grown.

"I buy King Edwards on the wholesale market, because there is a steady demand for them, but most of my potatoes are locally grown. I'm running Romana for a red and Estima for a white, and a grower from Newport brings them 100 bags at a time." 'The Country Kitchen' has about two hundred and eighty square feet of shop space, with half as much in the rear and the same again upstairs, for storage.

"I started off as a teenager when I left school and went to work for my Dad, who'd been in this business all his life. In ten years I've learned quite a bit. I'd been going down the market with my Dad for about twelve months when he had an accident in the van and was in hospital quite a while. So overnight I had to do it all myself. Now he's fully retired. Of course I still make mistakes, and anyone who says they don't make a mistake in this job is an idiot. But I'm not saying I make a lot of mistakes, because I can't afford to. There was an example just the other day. . . . I've been buying pumpkins for several years. They're all bought to scrape the middle out, to make pumpkin pie or whatever, and then the shell is used to make a jack-o'-lantern. So nobody wants to know about pumpkins the day after Hallowe'en. So I went down the market and priced up pumpkins, and when the gentleman said £2, what did I think? I thought it was £2 for a box of six, so I bought ten boxes! Sixty pumpkins! I should have known better, because when I went back to pay him I found it was £2 per pumpkin. I couldn't get out of it because they were short that particular day and it was too late when I went to square up with him. It was just one of those silly little mistakes. When I got them home, not only did nobody want to pay in excess of £2 to give me a little profit . . . I couldn't even sell them at £2. We ended up losing quite a lot of money.

"But that's only one blunder, usually I hit it right. I buy on Birmingham wholesale market, and sometimes on Wolverhampton. In summer plenty of local smallholders . . . market gardeners . . . come to me with their lettuce and tomatoes and other salads. I've a man who will be making holly wreaths soon who lives just three miles away. And I can get produce delivered, to top up with. If I know something is going to be short, like mushrooms at the moment with this sudden cold spell, then I'll order over the phone. But I don't like ordering blind. I like to buy in person because my first priority is quality. You look around at anything in this shop and I'd say that within reason it's about as good as you could buy anywhere. If I can get something that's quite a lot cheaper but still quite a smart commodity then I'll go for it, but I don't let the price factor rule the quality. You can see that I've got a class 1 Cox here which is an absolutely super sample, at 50 pence per pound. But I've also got a smaller, class 2 Cox which is still quite nice looking and will eat well, at 3 lb. for 78 pence. That's half the price.

"I leave home at quarter-past three in the morning to go to market with the van, three or four times a week as suits the trade. I deal with about fifteen firms, some more than others. You have your dog run. If a merchant is handling a particular mark of apples that I'm very happy with, I'll keep going back to him. But if the next consignment isn't up to scratch, I'll go elsewhere. It's a question of keeping your eyes open. A lot of the Evesham produce I buy from 'farmers'. Some of them are just people with a lorry, or perhaps they have some connection with a particular farm, and they go round the Vale collecting produce to sell on commission. One side of the market has a big lean-to and these chaps park under it and sit on the back of the lorry, selling their stuff right off it. I get back at about nine o'clock. My wife Annette and two other fulltime staff open the shop at half-past eight. It's long hours.

"I always try everything at the market, because quality isn't just the way something looks. But the public has to have an eye for a commodity. If you have a new line of bright, nicely coloured apples against, say, a Crispin that's a bit irregular in size and shape, or suspiciously yellow, they're going to buy the line that *looks* the part. But they might be coming back next week saying those apples were sour and flavourless, and straightaway you've got a problem — you've got to put it right or get down on your hands and knees and say you're awfully sorry, because you don't want to lose a customer. Sometimes you have to push something on a customer: 'Buy these instead, love, and you'll be pleased', and sure enough, they'll come back for more, because they can appreciate *real* quality.

"Sometimes people come in and ask me for organically grown lettuce. And older people, especially, tell me that the taste and quality of produce is not up to what it was years ago. Personally, I disagree. The reason we are having to use these pesticides is to keep the quality that will sell. If people bought a lettuce now and when they got it home found that it was full of greenfly, they'd bring it right back. But if enough people came and asked me to get organic lettuce. . . . Well, I don't know where I'd find it now, unless there was a small market gardener around who could assure me it was grown that way. . . . But if such lettuce were available and there was a demand, I'd certainly supply it. Somebody came in the other day and wanted six nectarines. They were very hard to find at this time of year, and I had to buy a whole box, but she got her nectarines.

"Our customers like the personal service. Whether they spend 10 pence or £5, they get a 'good morning' when they come in and a 'Thank you' when they go out, and any advice they might need in between. Supermarkets don't do that. There's an awful lot supermarkets can't do as well as we can. Fine Fare has moved into town since we've been here, but our takings haven't dropped. We keep taking more and more each year. As far as price goes, if they're buying 10 pallets and I'm buying 10 boxes, then they're obviously going to get a premium out on the price. But then they've got very different overheads. They have a large outlay. But if a supermarket invests in all these elaborate cold-chain storage facilities, it means they're reckoning on keeping their stuff. I don't have to worry about keeping it. I go and get my calabrese fresh every morning and sell it the same day and then go and get some more the next morning. And personally I think that's got to be a better way of doing it than buying a load and trying to keep it 'fresh' for a week.

"Of course when you get up at three o'clock and you walk outside the front door in the middle of winter and the windscreen on the van is frozen and you've got to drive all the way to Birmingham at thirty miles per hour because the fog is that thick you can't see far enough in front of you, and then you get down there and find that all the produce is frozen, so you have to buy the best you can and when you come back and you unload and you get all your shop set up and the first customer who comes through the door says, 'Haven't you got any better cabbage than that? It's all frosted!' . . . then you feel like jacking it all in.

"I don't think I ever would pack it in, though. It's the only trade I know, and it's the only trade that I think I could do as well as I do. I think this business has a good future, if the job's done properly.

"I'd like you to come and have a look at this shop at nine-thirty tomorrow morning, because this shop looks absolutely splendid. It really does look nice. And I'm proud to say that I will stand back outside and admire it. That's the time that you say to yourself, you're doing a good job, I'm sure a lot of people appreciate it, and you're glad to be in the trade."

INDEX

Index